The Young Specialist Looks At

Fungi

HANS HAAS

Translated and edited by Alfred Leutscher

Colour plates by Gabriele Gossner

Line drawings by Cynthia O'Brien

BURKE LONDON

First published in the English language June 1969
© BURKE PUBLISHING COMPANY LTD. 1969

Translated and adapted from *Pilze Mitteleuropas*
© Franckh'sche Verlagshandlung, W. Keller & Co. Stuttgart 1964 and 1966

ACKNOWLEDGEMENT

The cover photograph of *Nictalis asterophora* on *Russula nigricans* is reproduced by kind permission of Paul Popper

222   79409   7 Library Ed.
222   79414   3 Class Paperback Ed.

BURKE PUBLISHING COMPANY LIMITED
14 JOHN STREET   ★   LONDON, W.C.1

SET IN MONOPHOTO TIMES NEW ROMAN
MADE AND PRINTED IN GREAT BRITAIN
BY WILLIAM CLOWES AND SONS, LIMITED,
LONDON AND BECCLES

The Young Specialist Looks At
Fungi

Contents

FOREWORD   7

BIBLIOGRAPHY   8

GLOSSARY   9

ENGLISH NAMES OF FUNGI   11

PART ONE

The Biology and Life History of Fungi   15

Fairy Rings   20

Fungi in Sickness and Health   21

Fungi as Food   22

Truffles   23

Mushroom Growing   23

Poisonous Fungi   24

Mycorrhiza   26

Collecting and Identifying Fungi   27

Fungal Habitats   28

PART TWO

Colour Plates: 80 Fungi Illustrated and Fully Described   32

PART THREE

Key to the Agaricales   192

The Larger Fungi in Britain   194

Classified List   197

INDEX   236

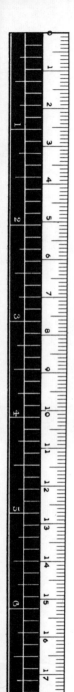

Scale showing inches and centimetres

Foreword

It is often assumed that mushrooms are edible and toadstools are poisonous. In fact, the so-called mushrooms which belong to the genus *Agaricus*, far from being the only edible fungi, have many rivals among their cousins the toadstools, and it should not come as a surprise that many of the species illustrated in this book are quite good to eat. **On the other hand, there are others which are poisonous, some even deadly. Care must always be taken, therefore, when collecting fungi for the table, only to gather those species on which a positive identification can be made.**

The reader will be helped in this identification by the eighty colour plates, in Part 2 of this book, of fungi which are to be found in Britain and western Europe. Each picture shows a background of beech-leaves, pine-needles, grass, etc., which suggests the kind of surroundings, or habitat, in which each species may be discovered. As a supplement to these, the book contains an extensive classified list of those species of the larger fungi which the amateur mycologist (as people who study fungi are called) may find in Britain during a foray.

Bibliography

Poisonous Fungi by John Ramsbottom (King Penguin 1945)

Edible Fungi by John Ramsbottom (King Penguin 1948)

Mushrooms and Toadstools by John Ramsbottom (Collins [New Naturalist] 1953)

A handbook of Larger Fungi by John Ramsbottom (British Museum [Natural History] 1965)

The Observer Book of Common Fungi by E. M. Wakefield (F. Warne 1954)

Wayside and Woodland Fungi by W. P. K. Findlay (F. Warne 1967)

Mushrooms by Pierre Montarnal (trans. by J. G. Barton) (Paul Hamlyn 1967)

Collins Guide to Mushrooms and Toadstools by Morten Lange and F. Bayard Hora (Collins, 3rd edition, 1967)

Glossary

Adnate broadly attached to the stem (of gills of an agaric toadstool) (*see* fig. 3D, p. 19)

Adnexed just reaching the stem (of gills of an agaric toadstool) (*see* fig. 3B, p. 19)

Annulus the ring round the stem of an agaric toadstool

Ascus the spore-sac of the Ascomycetes (*see* fig. 1D, p. 16)

Basidium the special reproductive cell in the Basidiomycetes which produces its spores externally (*see* fig. 1E, p. 16)

Boss a swelling (in the centre of the cap)

Capillitium a mass of thick-walled hyphae inside the fruit-body of Gasteromycetes (*e.g.*, puff-balls)

Conidium a spore produced asexually by the constriction of a hypha

Conidiophore a simple or branched fertile hypha on which conidia are produced (*see* fig. 1C, p. 16)

Cortina a web-like veil covering the gills in some agaric toadstools, especially in the genus *Cortinarius* (*see* fig. 3F, p. 19)

Crowded growing close together (of gills of an agaric toadstool)

Decurrent running down the stem (of gills of an agaric toadstool) (*see* fig. 3E, p. 19)

Dehisce to gape or burst open

Distant placed wide apart (of gills of an agaric toadstool)

Fibrillose covered with fine, hair-like fibres

Free not reaching the stem (of gills in an agaric toadstool) (*see* fig. 3A, p. 19)

Gleba the inner mass of Gasteromycetes (*e.g.*, puff-balls and earth-balls) and Tuberaceae (truffles)

Habitat the natural situation in which a plant or animal lives

Hymenium the fertile, spore-bearing layer in a fungus (*see* fig. 2A, p. 18)

Hypha a thread-like filament forming part of the vegetative mycelium of a fungus (*see* fig. 2, p. 18)

Mycelium the network of filamentous threads or hyphae (*q.v.*) which form the vegetative part of a fungus—*i.e.*, the "spawn" (*see* fig. 2, p. 18)

Mycology the scientific study of the Fungi

Mycorrhiza the association of a fungus mycelium with the roots of a higher plant, such as a tree

Peridioles the small spore-bodies, or "eggs", in the Bird's-nest fungi (Nidulariaceae)

Peridium   the skin forming the outer wall, particularly of the Gastero-
     mycetes

Perithecium   a flask-shaped structure containing asci in the Pyrenomycetes

Pileus   the umbrella-like cap in the Agaricales (*see* fig. 2, p. 18)

Remote   not reaching the stem (of gills of an agaric toadstool)

Rhizomorph   root-like strand of compressed mycelium

Rufous   a reddish or fox-like colour

Sessile   without a stem

Sinuate   wavy in outline, and curving suddenly before reaching the stem
     (of gills of an agaric toadstool) (*see* fig. 3C, p. 19)

Sporangium   the case or capsule which contains spores in the lower plants,
     including the Fungi (*see* fig. 1A, p. 16)

Spore   the reproductive cell in the lower plants, including the Fungi

Sporophore   the spore-bearing or fruiting portion of a fungus

Stipe   the stem of an agaric toadstool (*see* fig. 2, p. 18)

Striate   marked with lines, grooves or ridges

Veil   the outer envelope in the Agaricales inside which development of the
     fruit-body takes place (*see* fig. 2, p. 18)

Volva   the sack-like covering around the base of the stipe in some agarics
     (*e.g.*, *Amanita*) (*see* fig. 2, p. 18)

Zygospore   a resting spore resulting from the sexual union of two plants
     among the lower fungi (*i.e.*, Zygomycetes) (*see* fig. 1B, p. 16)

A List of some of the More Common English Names

Amethyst Toadstool (*Laccaria amethystina*) (*see plate 53*)
Beech Tuft (*Oudemansiella* (= *Armillaria*) *mucida*) (*see no. 51*)
Beef-steak (*Fistulina hepatica*) (*see no. 20*)
Birch Polypore (*Piptoporus betulinus*) (*see no. 10*)
Bird's-nest Fungi (*Nidulariaceae*) (*see no. 173*)
Blewitt or Blue-leg (*Rhodopaxillus personatus*) (*see plate 61*)
Blusher (*Amanita rubescens*) (*see plate 34*)
Buffcap (*Hygrophorus pratensis*) (*see no. 32*)
Chantarelle (*Cantharellus cibarius*) (*see plate 62*)
Cocoa-dust Toadstool (*Collybia maculata*) (*see no. 56*)
Common Inkcap (*Coprinus atramentarius*) (*see no. 92*)
Common Puff-ball (*Lycoperdon perlatum*) (*see plate 72*)
Common Stink-horn (*Phallus impudicus*) (*see no. 177*)
Cup Fungi (Pezizaceae) (*see nos. 185–6*)
Cultivated mushroom (*Agaricus bisporus* var. *albida*) (*see no. 85*)
Deathcap (*Amanita phalloides*) (*see plate 29*)
Destroying Angel (*Amanita virosa*) (*see plate 30*)
Devil's Boletus (*Boletus satanas*) (*see no. 150*)
Dog Stink-horn (*Mutinus caninus*) (*see no. 179*)
Dryad's Saddle (*Polyporus squamosus*) (*see no. 9*)
Dry-rot Fungus (*Merulius lacrymans*) (*see no. 3*)
Earth-balls (Sclerodermataceae) (*see nos. 165–6*)
Earth-stars (Geasteraceae) (*see nos. 171–2*)
Earth-tongues (Geoglossaceae) (*see no. 193*)
Fairy-ring Champignon (*Marasmius oreades*) (*see plate 52*)
False Blusher (*Amanita pantherina*) (*see plate 33*)
False Chantarelle (*Hygrophoropsis* (= *Clitocybe*) *aurantiaca*) (*see no. 35*)
False Deathcap (*Amanita citrina*) (*see plate 31*)
False Morels (*Helvella* spp.) (*see plate 77*)
Field Mushroom (*Agaricus campestris*) (*see no. 82*)

Fly Agaric (*Amanita muscaria*) (*see plate 32*)
Fool's Mushroom (*Amanita verna*) (*see no. 69*)
Giant Puff-ball (*Calvatia* (=*Lycoperdon*) *gigantea*) (*see no. 170*)
Grisette (*Amanitopsis vaginata*) (*see no. 76*)
Honey Fungus (*Armillariella* (=*Armillaria*) *mellea*) (*see plate 54*)
Horn-of-Plenty (*Cratarellus cornucopioides*) (*see plate 66*)
Horse Mushroom (*Agaricus arvensis*) (*see no. 83*)
Jelly fungi (Tremellaceae) (*see nos. 180–1*)
Lawyer's Wig (*Coprinus comatus*) (*see plate 28*)
Morels (Morchellaceae) (*see nos. 191–2*)
Orange-peel fungus (*Aleuria aurantia*) (*see no. 187*)
Oyster Mushroom (*Pleurotus ostreatus*) (*see no. 66*)
Pale Chantarelle (*Cantharellus cibarius*, var. *pallidus*) (*see plate 63*)
Panthercap (*Amanita pantherina*) (*see plate 33*)
Parasol Mushroom (*Lepiota procera*) (*see plate 35*)
Penny Bun (*Boletus edulis*) (*see plate 1*)
Puff-balls (Lycoperdaceae) (*see nos. 167–170*)
Razor-strop (*Piptoporus betulinus*) (*see no. 10*)
Rooting-shank (*Collybia radicata*) (*see no. 54*)
Round Morel (*Morchella esculenta*) (*see plate 78*)
Saffron Milkcap (*Lactarius deliciosus*) (*see plate 18*)
St. George's Mushroom (*Tricholoma gambosum*) (*see no. 36*)
Shaggy Inkcap (*Coprinus comatus*) (*see plate 28*)
Sickener (*Russula emetica*) (*see plate 27*)
Spindle-shank (*Collybia fusipes*) (*see no. 55*)
Stink-horns (Phallaceae) (*see nos. 177–9*)
Sulphur Tuft (*Nematoloma* (=*Hypholoma*) *fasciculare*) (*see plate 41*)
Tawny Grisette (*Amanitopsis fulva*) (*see no. 77*)
True Morels (*Morchella* spp.) (*see nos. 191–2*)
Truffles (Tuberaceae) (*see no. 195*)
Verdigris Toadstool (*Stropharia aeruginosa*) (*see no. 100*)
White Truffle (*Choiromyces meandriformis*) (*see no. 196*)
Wood Blewitt (*Rhodopaxillus* (=*Tricholoma*) *nudus*) (*see plate 60*)
Wood Hedgehog (*Hydnum repandum*) (*see plate 69*)
Wood Mushroom (*Agaricus silvicola*) (*see plate 38*)
Wood Woolly-foot (*Marasmius personatus*) *see no. 61*)
Woolly Milkcap (*Lactarius torminosus*) (*see plate 19*)
Yellow-staining Mushroom (*Agaricus xanthodermus*) (*see plate 39*)

Part One

The Biology and Life History of Fungi

Fungi are unlike any other form of plant life. They are unique both in their growth and appearance, and in their feeding habits. This is due to a total absence of chlorophyll, the essential green substance which, with the aid of sunlight, enables plants to manufacture their own food. Since this function is denied to fungi, they must obtain their food ready-made, as animals have to do. Some fungi, called *saprophytes*, obtain their food from dead material, upon which and inside which they grow. Some fungi, the true parasites, attack living things and may cause disease or death.

Together with bacteria, and a host of small ground-animals called "litter-fauna", the saprophytic fungi help to break down dead leaves and wood, and any other substance in which life has ended. In this way, they enrich the soil and help to build up humus.

Since no light is required for this kind of life, fungi can live in dark places where green plants would be at a disadvantage.

The larger kinds of fungi—those with which this book is concerned —are mainly concentrated in woods and grassland; it is these which are usually looked for during a foray.

The fruiting portion, called the *sporophore*, is the only part of the fungus which is visible. It takes the form of a toadstool, bracket, puff-ball, stink-horn, etc. These appear mainly in the mild days of autumn. Hidden in the leaf-mould or dead wood is the permanent plant, a thread-like structure which feeds on the material around it. This is the *mycelium*, or "spawn" as mushroom-growers call it. It continues to live and spread, year after year, until the food-supply is exhausted. This explains why a certain kind of toadstool or bracket turns up in the same spot every autumn.

Fungi vary widely in their shape and size, but fall naturally into three major groups: Phycomycetes, Ascomycetes and Basidiomycetes. These are distinguished by the structure of the reproductive region from which the spores are produced. (In addition there are the Fungi Imperfecti which lack a complete life-cycle.)

In the Phycomycetes group, most of the numerous species are composed of filament-like threads, microscopic in size, and resembling thread-like algae except for their lack of chlorophyll. Some of these, popularly called moulds, feed on non-living organic substances such as food, clothing and paper. Others attack living things—as, for example, mildews which attack plants, and ringworm which attack animals. A common form of reproduction is by means of a *sporangium*, which produces spores. This is an asexual method from a single plant. Some phycomycetes (the zygomycetes) reproduce sexually by *zygospore* formation (figure 1A and B).

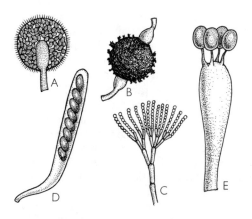

Fig. 1.   Reproductive organs

A. a sporangium (asexual organ) in *Mucor*—Phyco-
mycetes
B. a zygospore (sexual spore) in *Mucor*
C. a conidiophore in *Penicillium*—Fungi Imperfecti
D. an ascus with 8 spores—Ascomycetes
E. a basidium with 4 spores—Basidiomycetes

Many ascomycetes are microfungi; others are visible to the naked eye; some—the truffles—grow below ground. Ascomycetes include yeasts, morels, cup fungi, and the parasitic ergot. The spores are produced in batches of eight inside a flask-shaped container, called an *ascus* (figure 1D).

Amongst the third group, the Basidiomycetes, are found the familiar toadstools, mushrooms, brackets, fairy-clubs, earth-balls and many others. In these species the spores are produced in fours, and are borne on a club-shaped structure called a *basidium* (figure 1E).

A fungal spore is a simple structure, about one thousandth of an inch in size. It does not germinate in the sense that a flower's seed does, as there is no embryo. The spore's contents are mostly colourless and contain a little oil, probably as a food reserve. A vast number of spores may be liberated by a single sporophore. A Field Mushroom with a four-inch cap is estimated to release something like one hundred thousand spores an hour, night and day, for four days. Britain's largest fungus, the Giant Puff-ball, may liberate some seven billion spores during its short fruiting period! The fact that they are so light and so numerous means that they are efficiently distributed, and few organic substances are free of them.

Should a spore settle upon a suitable substance, it will, under the proper conditions of temperature and humidity, proceed to grow by sending out a pale thread, called a *hypha*. This can branch into many hyphae, so as to form a dense weft of mycelium resembling a mass of cotton wool. This is the actual fungus plant to be found buried in the leaf-litter or dead wood. From time to time, mainly in autumn, the mycelium sends out a number of fruit-bodies. These are what the fungus collector gathers on his outings.

Perhaps the simplest way of illustrating how a fruit-body grows, is to take an example from the genus of toadstool called *Amanita*. These form part of the family Agaricaceae, which are easily recognizable because of their gills. The species chosen is the familiar Fly Agaric, *Amanita muscaria* (*see* plate 32). Some time during the mild days of September or October, the agaric sends up its sporophores (figure 2). This starts as a tiny, compact underground mass of interwoven hyphae, in the shape of a small white ball. The ball is enclosed in an outer skin, called the "universal veil". As it breaks out of the soil, it enlarges into the familiar button shape. Gills develop inside it. As the first stage in the gills' development, a dark staining layer of hyphae, called the *hymenophore*, appears in the form of a ring inside the button. Then, below this, a cavity is formed, which is filled by a special layer called the *hymenium*. This is the ripe fruiting layer on which the gills will develop—since this species is a gilled fungus. Then, once the gills have grown, the basidia, with their four spores each, will finally appear. (Not all toadstools have gills. For example, the Hydnaceae (p. 198) produce their spores on spines under the cap; the Boletaceae (p. 223) have them inside tubes which open at the surface as pores. In the Clavariaceae (p. 202) spores are produced on simple or branched sporophores.)

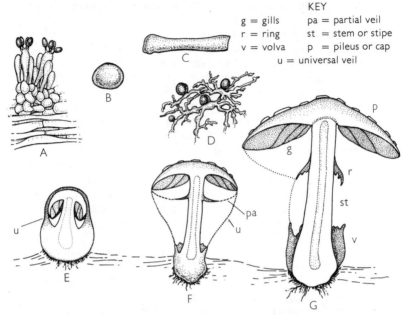

KEY

g = gills   pa = partial veil
r = ring   st = stem or stipe
v = volva   p = pileus or cap
       u = universal veil

Fig. 2.   Development of a toadstool (*Amanita*)

A. the fruiting layer (hymenium) with basidia bearing spores
B. a single spore
C. germinating spore
D. branched hyphae forming a mycelium—*i.e.*, spawn—with young fruit-bodies
E. a young fruit-body (button stage)
F. a young toadstool emerging from the universal veil
G. fully grown *Amanita* toadstool

Note: A, B, C and D are all greatly enlarged

As the button grows, it expands into a club shape. The lower part will form the stem, or *stipe*, and the club portion the cap, or *pileus*. As the cap expands, it opens like an umbrella and, in so doing, breaks through the universal veil. Parts of the veil break into small patches and are carried onto the top of the cap. These white spots show up clearly on the red cap. The remainder stays wrapped around the stem base, and is called the *volva*. The expanding cap also tears away from the stem. At first it is joined to the stem by another skin, known as the "partial veil" which hides the gills. Finally this is torn and remains as a ring round the stem (figure 2). In some genera, such as

Cortinarius (*see* p. 217), this veil is very fine and lace-like, and can be seen stretched across the gills from the stem to the edge of the cap, especially in young specimens (figure 3F).

When ripe, the spores are liberated from their basidia by an ingenious method. A drop of water appears upon the spore. As the drop dries it sets up surface tensions, causing the spore to be shot off the basidium, into the space between the gills. These hang downwards, and must be aligned at right angles to the direction of gravity; otherwise the spores cannot escape. Consequently, a toadstool, in whatever position it grows, will always straighten its cap so that it is parallel to the earth. This can even happen overnight with specimens laid down sideways in the collector's basket, where the cap of a toadstool will turn into the horizontal position. The spore falls in the direction of gravity, and as it escapes it is carried on the free air to its destination.

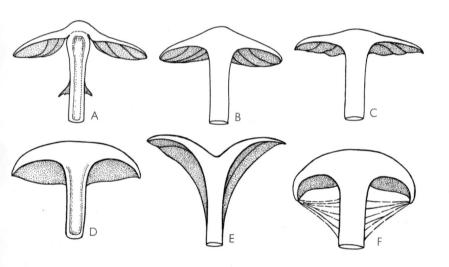

Fig. 3. Gill attachment in the Agaricales
A. free—gills do not reach stem
B. adnexed—gills just reach stem
C. sinuate—gills have wavy outline and curve suddenly before reaching stem
D. adnate—gills broadly attached to stem
E. decurrent—gills run down stem
F. a veil or cortina (*e.g.*, Cortinariaceae)

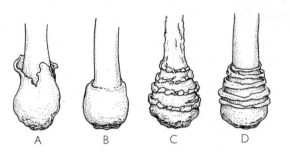

Fig. 4. Structure of the volva in different species of *Amanita*
A. *Amanita phalloides* C. *Amanita muscaria*
B. *Amanita citrina* D. *Amanita pantherina*

The *Amanita* group are sometimes described as the perfect example of toadstools, since they carry all the essential parts used in identification (*see* p. 210). Not all agarics have both ring and volva—some have one or the other, and some neither.

Fairy Rings

There was a time when all kinds of fanciful explanations were given for the origin of "fairy rings". People believed that they were made by fairies, and that anyone who stepped inside the circle came under the spell of the fairy folk. In France, nobody would enter a fairy ring because it was said to be inhabited by enormous toads, believed to be venemous. Others thought the rings were the work of moles, ants or foxes; of lightning, or whirlwinds, or that they were caused by animal urine.

It was not until 1790 that the truth finally became known. In fact, fairy rings are caused by the circular growth of the mycelium of certain small toadstools, especially *Marasmius oreades*, the Fairy-ring Champignon (*see* plate 52). The ring commences with the radial spread of the mycelium which advances outwards at a rate of some 6 to 12 inches a year. The ring is composed of three zones, and the variation in the colour and texture of the grass in the three zones is due to the activities of the fungus.

This is what happens: The fungus feeds off dead matter in the soil. The carbohydrates are used up, but some of the protein matter is converted into ammonia. The ammonia forms salts and some of these, such as nitrates, are changed through bacterial action into nitrogen compounds. These fertilize the soil and stimulate growth. The result is a zone of rich grass along the outer border of the mycelium ring. Behind this is a zone of dried-out grass, where the mycelium is well developed, possibly reducing aeration and using up the water supply. It is here that fruit-bodies are formed. In the centre of the ring is a third zone, also of rich grass, where the mycelium has rotted away, and the nitrogen which is released acts as food for the grass.

Fungi in Sickness and Health

Fungal attacks on the human person are not normally serious if treated in time, and are usually due to neglect and poor hygiene. A fungal skin complaint known as ringworm may be picked up from infected cattle and other animals, and can cause hair to fall.

Another complaint, known as "athlete's foot", occurs where the fungus settles on the skin in places which are likely to perspire, such as between the toes. Athletes, who perspire freely as the result of their exertions, are especially prone to this disease.

It has recently been found that moulds which grow on food can produce poisonous side-effects which are thought to cause cancer of the liver. Even more unusual is the arsenical poisoning resulting from the gases given off by moulds which grow on damp walls and coloured wallpaper, where the latter contain traces of arsenic.

Ergotism, known for centuries as "St. Anthony's Fire", can be a serious complaint both in animals and in man, and is caused by eating grain or bread infected with an ascomycete *Claviceps purpurea*. This is the fungus ergot which parasitizes grasses and cereals, especially rye.

On the credit side, must be mentioned the "wonder drugs" which are extracted from fungi, notably penicillin. Sir Alexander Fleming's discovery in 1929, when he noticed the inhibitory effect of the mould *Penicillium notatum* on a culture of bacteria, was a milestone in medical science, leading to the use of antibiotics in medicine. Several other fungal drugs in general medical use today are extracted from ergot. The destructive effects of this fungus have already been described, but it can also be most useful as a muscle stimulant. It assists in childbirth and aids the mother in the contractions of the uterus during delivery. It is interesting to note in this connection that down the ages certain fields in farming districts have been avoided as places of evil, since so many still births have occurred amongst sheep and cattle. This, in all probability, has been due to the presence of *Claviceps* on the grass, resulting in premature births among the animals.

Fungi as Food

Because of our somewhat suspicious natures, most English-speaking people treat fungi as plants to be avoided—with the exception, that is, of the cultivated mushroom sold in the shops all the year round (*see* p. 212). Compared with the wild mushroom, this cultivated cousin has an insipid taste. There are also a number of toadstools which are not only edible but just as tasty as any mushroom, contrary to the ancient belief that "mushrooms are edible and toadstools are poisonous". In fact, mycologists often use the two words interchangeably. However, it is generally agreed that the word "mushroom" should be confined to the genus *Agaricus*. Of these the Field Mushroom (*A. campestris*) and Horse Mushroom (*A. arvensis*) are best known. The cultivated mushroom is believed to come from another species, *A. bisporus* (*see* no. 84).

Other mushrooms which are good to eat, from other genera, are the so-called St. George's Mushroom (*Tricholoma gambosum*) which fruits in spring, the Parasol (*Lepiota procera*) and the Oyster Mushroom (*Pleurotus ostreatus*). Amongst the toadstools, a particular favourite is the late autumn-growing Blewitt, and its cousin the Wood Blewitt, *Rhodopaxillus* (=*Tricholoma*) *personatus* and *R. nudus*. Puff-balls, especially young specimens which are still white inside, are delicious cooked in butter. The Giant Puff-ball, *Calvatia giganteum*, tastes particularly good; so does the Shaggy Inkcap, *Coprinus comatus*.

Further examples of good edible fungi are the Saffron Milkcap, *Lactarius deliciosus*, the chantarelle, *Cantharellus cibarius*, and the famous Penny Bun, *Boletus edulis*. The last is widely collected in France (where it is known as "cèpe") and Germany (*see* plate 1). These fungi may be dried and used as flavouring in soups. Many of the smaller toadstools, such as the Fairy-ring Champignon (*Marasmius oreades*) can be dried and stored, then later added for

22

extra flavour to such dishes as casseroles, stews, eggs-on-toast, cheese dishes and various sauces.

People on the Continent use a wide variety of fungi in their cooking, especially in France where the preparation of toadstools has been brought to perfection.

Truffles

The famous truffle is still the most prized and the most eagerly sought-after in France. Unfortunately, this fungus is now a rarity in Britain; the art of finding it has been lost almost entirely. Truffles are ascomycetes which grow below ground (genus *Tuber*), and can be found only by their scent. Dogs, usually of a poodle breed, are trained to sniff them out. Occasionally pigs can be made to uproot them. In Britain, truffles grow mainly in calcareous soil under beech trees. In France they occur under the evergreen oaks, and the truffle has been cultivated in France and other countries by planting young oaks and then inoculating the ground with soil from a truffle area. The species of truffle found in Britain are given on pages 233–4.

Mushroom Growing

Although the common mushroom is popular and widely cultivated, neither it nor its two wild cousins, the Field and Horse Mushrooms, appear to have been used much in classical times. The actual culture of mushrooms is first recorded in France as late as the beginning of the eighteenth century and appears to have differed little from present-day methods of cultivation. Firstly the mycelium, or "spawn", must be collected. It is found in rich pasture-land where horses or cattle are kept. Portions of this virgin spawn are injected into brick-like slabs consisting of a mixture of moistened stable manure and loam. The slabs are made by pressing the mixture into a rectangular mould about the size of a house-brick. Then, into each slab of the dried mixture, or between the layers, are pressed small pieces of the virgin spawn, so that the slabs are thoroughly inoculated with mycelia. The slabs can be stored and used up to a year later.

The best compost for growing mushrooms is a stable manure from grain-fed animals. It should be fresh and mixed with plenty of bedding straw. This is piled into heaps, turned over now and then, and kept wet so that it ferments. The process of fermentation should take some days at a temperature of about 49° C., or 120° F., until the manure smell disappears and the straw becomes pliable. The purpose of fermentation is to render the mass chemically suitable for the mycelium to feed on, so that it can extract the soluble organic products as well as the salts from the straw and manure.

The compost is either pressed down with a shovel into beds about one foot deep, or put into boxes. It is then "spawned" by breaking egg-sized pieces off the brick-like slabs and burying these in every square foot of the compost. When the mycelium begins to take a hold, a rich layer of loam is placed on top. Then, once the mushrooms start to appear, they should be given a regular sprinkling to keep the loam moist. A good average temperature is 23° C. (75° F.).

It is not essential to grow mushrooms in darkness, although this is the usual practice. In fact, the best situations for growing mushrooms are such places as caves and underground passages like the famous catacombs underneath Paris, which are dark anyway. The important thing is to avoid draughts, and to maintain the proper temperature level.

Poisonous Fungi

There are few deadly-poisonous fungi in Britain, and one of these, the Deathcap, *Amanita phalloides* (plate 29) probably accounts for ninety per cent of deaths from fungus poisoning. The fungus collector who wishes to experiment with toadstool dishes should get to know this species and its near allies absolutely thoroughly. It is only by knowing the characteristics displayed by each species that he will be able to identify them with certainty. There is *no* rule-of-thumb test which is safe. Contrary to popular belief, it is impossible to tell from such signs as the tarnishing of a silver spoon or coin, or from examining the smell, the skin peeling, whether it is coloured white or grows in grass, for all these can apply to both edible and poisonous kinds.

The Destroying Angel (*A. virosa*), for all its innocent whiteness, is a killer (plate 30). So is the Fool's Mushroom (*A. verna*). These three—*A. phalloides*, *A. virosa* and *A. verna*—should be treated with every respect. A useful reminder is that they have white gills, whereas the genus *Agaricus*—except in the extremely small button stage, where it is still wrapped in its universal veil—has pink gills which darken with age.

Symptoms of poisoning are due to (a) an alcohol-soluble toxin, called amanitin, which is not affected by heat, drying or digestive juices; and (b) phalloidin, a polypeptide which acts rapidly but which can be destroyed by heat. The former poison appears to be the chief cause of death. Another ingredient called phallin, once thought to be the killing agent, is now known to attack the red blood corpuscles, but is readily destroyed by the stomach juices.

Symptoms follow a characteristic course. After an initial incubation period which lasts for ten hours or so, there is a sudden onset of stomach cramps, vomiting, thirst and diarrhoea, also passing of blood and mucus. This lasts for about two days and is then followed by another quiet interval. This is the danger period. If only a tiny portion of fungus has been eaten there may be recovery, although this will be very slow. Otherwise all the symptoms return with greater severity, paralysis occurs and leads to final collapse. Death then follows on the third or fourth day.

Various methods of cure have been tried with some success. The Pasteur Institute in France has produced an anti-phalloidian serum from immunized horses. A 40 per cent dose is injected intravenously. A 25 per cent glucose injection has also been tried with some relief, since the poison causes a shortage of sugar in the blood, a condition called hypoglycaemia.

A 20 per cent salt solution is also given, either by mouth or injection, in 20 c.c. doses.

One old method of cure was based on the knowledge that rabbits can eat poisonous fungi without harm, suggesting that the rabbit's stomach-lining might contain a substance which neutralizes the poison. The patient was given chopped-up stomach of rabbit to eat, mixed with brains. The former neutralizes the poison which affects the liver, and the latter the neurotoxin which attacks the nerves and brings on the paralysis.

The Fly Agaric *A. muscaria* (plate 32) is another poisonous toadstool; but the results of eating this are not usually fatal. The symptoms in this case are less severe, but colic, vomiting and diarrhoea may occur. In some people, excitement and hallucinations resembling drunkenness may occur also, leading to a comatose condition. Delirium, convulsions, prostration and sleepiness have been observed; although these symptoms are somewhat alarming, there is a quick recovery from *A. muscaria* poisoning.

This toadstool has been used by northern Asiatic tribes to produce

25

deliberate intoxication in religious ceremonies, or merely to produce high spirits. It is said that the ancient Viking raiders would make themselves fighting mad, or berserk, by consuming great quantities of *A. muscaria*.

The substances involved are alkaloids. Of these, it is muscaridine which affects the nervous system and choline which causes the gastric disorders. The *Fly Agaric* got its name from the old country practice of using pieces of the fungus broken up in milk as a fly killer.

Other fungi which need to be treated with caution are listed below. All of them, including the species above, are listed in the classified section which starts on page 194.

Amanita pantherina – the Panthercap (plate 33)
Inocybe fastigiata – (plate 45)
Inocybe geophylla (plate 46)
Inocybe patouillardii – the Red-staining Inocybe (*see* page 218)
Rhodophyllus sinuatus (= *Entoloma lividum)* – the Leaden Entoloma (plate 48)
Clitocybe dealbata (plate 51)
Tricholoma pardinum (plate 57)
Clavaria pallida (plate 68)
Lactarius torminosus – the Woolly Milkcap (plate 19)
Russula emetica – the Sickener (plate 27)
Agaricus xanthordermus – the Yellow-staining Mushroom (plate 39)
Boletus satanas – the Devil's Boletus (no. 150)

The fungus-eater would be wise not to experiment with his finds without a positive identification. Of the edible kinds, only fresh specimens should be eaten and even these should not be eaten in too great a quantity. In actual fact, the food value of fungi is not very high: up to 90 per cent of the contents is water. Of the remainder, about 5 per cent is protein with some traces of carbohydrates, a little mineral salt and some vitamins.

Mycorrhiza

During a field ramble, you may notice that certain toadstools and other fungi grow in close association with certain species of trees. In many cases it is known that the roots of the tree are closely bound up with the mycelium of the fungus. Its hyphae completely envelop the tree's rootlets; to the extent that the tree may even lack any root-hairs of its own through which it would normally take in its food in water solution. This presupposes that the fungal hyphae are taking over the function of the hairs. Such a close link between tree and fungus is called *mycorrhiza*, and is usually considered to be a symbiotic relationship—i.e., the intimate joining together of two distinct organisms for mutual benefit. Water and inorganic salts pass from the hyphae into the tree, and the fungus possibly receives some nourishment from the tree in exchange.

Some plants are unable to exist without the co-operation of a fungus. This is especially true of orchids and heathers. Orchid seeds will not germinate unless the fungus is present.

Collecting and Identifying Fungi

In preparing for an outing to collect fungi, certain articles will be needed. The most obvious of these is a suitable collecting receptacle: a stiff basket or, if this is not available, a garden trug or cardboard box. This should be lined with moss, picked up on the way, so that the specimen can be placed on a soft, mossy bed, with less risk of damage. (The moss can be used later in the creation of a toadstool garden.) Small and delicate specimens can be placed between pads of cotton wool inside a small tube, or pinned to cork inside a small tin or box, as for insects.

If a serious study is to be made of the collection, then a field guide, a hand-lens and a notebook and camera will be useful. The notebook will record the date, weather, habitat and surrounding vegetation. The last is important, as many fungi have special habitats (*see* Fungal Habitats p. 28). Photographs may be taken on the foray, or colour drawings made later at home. Spore prints are also useful for recording the colour of the spores of agaric toadstools. In order to make a spore print, cut away the cap from the stem, and lay this, gills down, on a sheet of cardboard or stiff paper (white for dark-spored species —e.g., *Agaricus*; and coloured for white species—e.g., *Amanita*). Cover the specimen with a bowl so that the air is still, and leave overnight. The gills should have dropped enough spores for the print to be complete by the next morning. If covered with a cellophane sheet it should last for years.

The actual preservation of fungi will depend upon the species and its texture. Many of the brackets (Polyporaceae) which are hard, woody or corky, can be dried and stored in boxes. Of the softer kinds, such as toadstools, a thin section can be cut longitudinally, placed between drying paper and pressed as for flowering plants. Some good results have been obtained lately by freeze-drying specimens in a vacuum chamber. Fungi prepared in this way are on display in the Botany gallery of the Natural History Museum, in South Kensington, London.

Apart from toadstools—i.e., all those species which have a cap and stem—all other fungi—brackets, stink-horns, puff-balls cup fungi, fairy clubs, morels, etc.—should be easy to recognize. When examining toadstools, there are a number of ways of identifying specimens, by observing the features or answering the questions listed below (*see also* the Key on pages 192–3):

1. Is it gilled (Agaric) pored (Boletaceae) toothed or spined (Hydnaceae)?
2. Note the size, texture and toughness of the flesh and skin.
3. Is there a ring and/or volva present?
4. Is there a change of colour when the flesh is pressed (as in the pores, in some Boleti)?
5. How are the gills attached to Agaric toadstools? (*See* figure 3.)
6. Does it exude milk (as in *Lactarius*)?
7. Is there a characteristic smell or taste?
8. Note the colour, which may be a strong feature. (**Warning: some toadstools change colour with age!**)
9. Microscopic characters: some species have spores with a peculiar shape and surface, others have hyphae under the surface of the cap which may be cellular or filamentous, and some display cystidia (sterile cells) between the basidia in the hymenial layer.

Fungal Habitats

As already stated, fungi rely on other organisms for their food. Consequently their appearance in any specific locality depends more on the local type of vegetation present than on the kind of soil. For example, in a conifer wood, the composition of the humus formed by the fallen needles and branches is quite different from the leaf-mould of a beech or oak wood. In each kind of wood different species of fungus will be found. Other species are even more closely tied to a certain kind of tree by the mycorrhiza.

Any alteration in the landscape, such as planting a conifer wood on open pasture-land, will in time alter the whole complex of fungi, from those normally inhabiting grassland to those found under needle trees. It is even possible to find specimens peculiar to conifer woods growing under a single pine tree in the middle of an oak wood. Other exceptional conditions, such as those found on a patch of burnt ground, a well-manured area or a dung-heap, will also attract specific types of fungus.

Some of the more usual surroundings in which fungi are found, together with some of the typical species found in them, are listed here.

1. Woodland
   (a) Coniferous wood: *Xerocomus badius* and *bovinus*; *Hydnum repandum*; *Lactarius deliciosus*; *Russula drimeia* and *Rhodopaxillus nudus*.
   (b) Deciduous wood: *Amanita phalloides*; *Boletus edulis* and *B. scaber*; *Cantharellus cibarius*; *Russula emetica*; *Cratarellus cornucopioides*; *Lycoperdon perlatum*; *Scleroderma aurantium*.
   (c) Mixed wood: *Amanita rubescens* and *A. citrina*; *Amanitopsis vaginata* and *A. fulva*; *Collybia maculata*; *Laccaria amethystina*; *Paxillus involutus*; *Russula ochroleuca*.

2. Pastures and Lawns
Agaricus arvensis and *A. campestris*; *Lepiota procera*; *Marasmius oreades*; *Tricholoma gambosum*.

3. Burnt Ground
Pholiota carbonaria; *Lactarius quietus*; and various cup fungi.

4. Animal Dung
Panaeolus spp. and *Coprinus* spp.; *Bolbitius vitellinus*.

5. Below Ground
Truffles (Tuber), mainly under beech on chalky soil in Britain, under evergreen oaks in France (*see* nos. 195 and 196).

6. Timber
Mainly bracket fungi (Polyporaceae)
   (a) Standing trees: *Polyporus sulphureus* on conifer and *Piptoporus betulinus* on birch; *Ganoderma applanatum* and

Armillariella mellea on beech; *Fistulina hepatica* on oak; *P. squamosus* on elm; *Auricularia auriculajudae* on elder.

(b) Fallen or worked timber; tree-stumps: *Stereum purpureum*; *Trametes gibbosa*; *Polystictus versicolor* on beech; *Auricularia mesenterica* on elm; *Trametes quercina, Stereum hirsutum* and *Bulgaria inquinans* on oak. *Merulius lacrymans* on timber in buildings and mines. *Hypholoma fasciculare* and *Pholiota squarrosa* on tree-stumps.

Part Two

Boletus edulis Penny Bun
(family Boletaceae) **Edible**

Description

The cap of this much sought-after and commercialized edible species is white at first, soon turning brown, and finally a liver-, nut- or blackish-brown. It may also turn grey or reddish. It can reach 8 in. in width, or even as much as 16 in. The smooth or finely wrinkled skin becomes a little slippery after rain. Under the cap is a separate layer of very fine tubes. These and their pores enlarge with growth, turning yellow to olive-green. The firm and rounded white stalk grows to a length of up to 12 in. and may turn brownish, in which case the fine network of white lines shows up clearly on the upper stalk. The flesh is a clear white, but may have a reddish tinge just under the skin of the cap. It has an agreeable, nutty taste, and does not stain blue.

Habitat

This species will only bear fruit when it grows near woodland trees with which it associates through its mycorrhiza (*see* p. 26). The trees are mainly conifers (pine and fir), beech, oak and birch. Sometimes this boletus can be found in masses in young thickets; it is also found under old trees in open woods, and even under isolated trees in parks. Large numbers appear only every few years. After dry or cool weather, the toadstools are usually full of maggot holes. Sandy, especially loamy, soil, with acid leaf-mould, is favoured by this species, which does not grow well in pure chalky ground. (The well-known Fly Agaric, with its white-spotted, gleaming-red cap, is often seen by the collector growing near by, since it requires the same kind of conditions as *B. edulis*.) The Penny Bun appears from May onwards.

Edibility

This species is one of the most popular edible fungi, and is widely used in cooking. It has a very fine flavour, and it grows abundantly. These two factors have made it one of the most important marketable products of its kind in Europe.

Similar species

It may occur with the very bitter tasting but harmless *Tylopilus* (= *Boletus*) *felleus* (*see* plate 2). In this species the cap and stem are more olive-brown, the tubes a flesh-pink and the network on the stem much coarser. Tasting the smallest fragment of the bitter flesh will immediately identify it.

Note: a number of the pored toadstools have blood-red pores. Their flesh quickly stains blue when cut. Such species should be avoided by the collector —i.e. *B. luridus*, *B. erythropus*, *B. satanas* and *B. purpureus*.

PLATE 1

Tylopilus (= Boletus) felleus
(family Boletaceae)   ☹ **Inedible**

Description

In its early stages of growth, this species can easily be mistaken for *Boletus edulis* (*see* plate 1). However, the brown of the cap is tinged with olive, and the stem has a coarser reticulated surface which is never found on *B. edulis*. At first egg-shaped, it later becomes tall and slender. There are, however, stubby forms to be found. The tubes gradually change from white to flesh-pink. As they lengthen, they give the underside of the cap a curved shape. The pure white inner flesh looks very appetizing, and when cut or bruised hardly stains at all. However, the taste is generally so bitter that it can be detected in the smallest sample. Nevertheless it is possible to find the odd specimen which tastes quite mild.

Habitat

This species appears from summer onwards, mainly in sandy, gravelly or peaty soils under conifers. It is not usually found in deciduous woods. In some years, depending on the weather conditions, it turns up in large numbers when the Penny Bun is scarce. In other years the reverse situation occurs. There is a similar situation with the Chantarelle *Cantharellus cibarius* and the False Chantarelle *Hygrophoropsis* (= *Clitocybe*) *aurantiaca*.

Edibility

Because of its extreme bitterness this is not an edible fungus. Even the odd specimen, mixed in with other, edible fungi, can be detected by its acrid taste.

Similar species

A young specimen, with its white pores and bulbous stem, could be mistaken for a Penny Bun, but is unmistakable when tasted or bruised. Other visible characters to note are the yellowish tubes which turn olive-green; the flesh colour which does not stain, and the fine meshwork pattern on the upper stem. The colour change in the tubes when broken is sufficient evidence to identify the genus *Tylopilus*.

Occasionally a species *Xerocomus* (= *Boletus*) *badius* will share a similar habitat (*see* plate 6). However, the colour difference of the tubes and the condition of the flesh when broken should distinguish these two species with little trouble.

The olive cap of *X. submentosus* may occasionally resemble *felleus*, but it differs otherwise in having yellow tubes and a thin stem, usually without network markings which occur, if at all, near the apex.

PLATE 2

Boletus calopus
(family Boletaceae)   **Inedible**

Description

This boletus has a reddish stem streaked with yellow, especially when young, and bears a resemblance to *B. luridus* (plate 4). The colour of the cap is grey-white tinged with olive-brown. The tubes are yellow, later turning olive-green. This toadstool is sensitive to the touch and turns blue when pressed, especially when cut. The intense blue spreads slowly from the damaged part. Because of its extremely bitter flavour this is not an edible species.

Habitat

B. calopus is found mainly in coniferous woods where it appears in summer in clusters. Some specimens grow to a large size; other, smaller specimens, with thinner stems, may be found among deciduous trees. Its favoured soil is sandy. Those found on chalky ground may look so dissimilar as to be mistaken for another species.

Edibility

This is not an edible species. However, although often spoken of as having poisonous properties, this has not been proved.

Similar species

Dingy specimens may be mistaken for *Xerocomus chrysenteron* (plate 7) but the latter has no network pattern on its stem. Larger and paler forms sometimes resemble *B. luridus* (plate 4), except that in *B. calopus* there is a total absence of red pores. In beech woods, specimens of *B. satanas* (no. 150) may be found with a similar coloured cap, but the thick stem and red pores will tell them apart.

PLATE 3

Boletus luridus
(family Boletaceae)   🙂 **Most specimens edible if well cooked**

Description
The cushion-shaped cap is a brown-olive, to olive-ochre tinted with purple, and is covered with a fine olive-yellow downy lustre. Tubes are yellow to olive-green with orange pores turning through chrome-yellow to a dull purple. When bruised, they turn deep blue, and eventually the redness disappears. As in all boleti, the tubes separate easily from the cap to reveal a chrome-yellow to orange-red base. The stem is a deep red on a pale earth-coloured background on which there is a network of red lines. The flesh of the cap is pale red, the stem yellow but turning dark red in seconds when cut, then blue. There is a sharp, sour smell, but this is not very noticeable. Apart from an occasional poorly developed specimen, many are thick-stemmed and reach a cap-width of 8 in.

Habitat
This species, which has many degenerate forms, appears in June on chalky soil in deciduous woods, rarely among conifers, and avoids acid soils. It can be found along woodland paths, or even in grass under a solitary birch or beech tree.

Edibility
Many toadstool gatherers, especially in France, recommend this species. Insufficiently cooked or eaten raw, however, it can certainly cause unpleasant gastric upsets. One should always avoid those boleti with a veined stem.

Similar species
The amateur collector usually terms all boleti which have red pores and turn blue when bruised "Satan toadstools". The true species, *B. satanas* has a pale grey or grey-green cap, and has glowing blood-red pores. The stem is a clear red with yellow network, and is conspicuous by its rounded dumpy shape. A smell of decay is typical in young specimens. The yellowish flesh turns a pale blue when cut. It is confined to chalk and grows under deciduous trees, mainly copper-beech. *B. satanas* is poisonous, but not deadly. On the other hand, the species *B. erythropus* is a good eater. It has a dark brown cap, carmine pores, a speckled stem and clear yellow flesh turning a deep blue. It grows on sand among conifers. The rare *B. purpureus* has red pores, a reticulated stem and yellow flesh.

PLATE 4

Leccinium (= Boletus) rufescens
(family Boletaceae)   **Edible**

Description
The pleasing orange to brick-red cap of this boletus can reach a width of
10 in. The felt-like skin has a small fringe which overlaps the edge. In the
early stages of growth, this skin is joined to the stem to cover the pores;
later it is torn free. The tubes are greyish; the pores are a little darker, turning olive when bruised. The length of the tubes can reach $1\frac{1}{2}$ in. in fully
grown specimens. The very striking and solid stem is long and has a rough,
black, scaly surface. In young specimens, the small cap is attached to the
stem by a narrow skin, and looks like a rounded lid. The ground colouring
of the stem often shows a number of dark blue-green spots which may
penetrate into the flesh. Whereas the stem remains firm, the cap is much
softer and finally collapses. When cut, the flesh is white, then turns through
dirty lilac to wine red.

Habitat
This species appears in summer and autumn in many forms. It takes on
the beautiful brick-red colouring shown in the illustration usually when
growing under birch and sometimes with conifers in sandy or loamy soils,
especially in company with heather and bilberries. A browner variety, var.
aurantiacum, has reddish patches on the stem and violet-grey flesh colours.
It is found growing under aspen. Under birch may be found a hazel-brown
variety which could be confused with *L. scabrum* (*see* no. 154).

Edibility
This is good to eat, as long as the flesh is firm. It is attacked far less by
maggots than the Penny Bun. It should be noted that, when cooked, the
flesh turns blackish, but this is not cause for alarm.

Similar species
L. rufescens is so characteristic, especially in its brick-red form, that it
should be easy to recognize. All closely related species are edible as well.
L. scabrum, to be found under birch, has a slender stem and a soft, rather
thin cap, coloured a grey-brown, with a smooth, viscid (sticky) skin. The
flesh does not stain. Another species, *L. carpini*, resembles these two in form
and colour. It is to be found in beechwoods. Its flesh stains a deep violet.

PLATE 5

Xerocomus (= Boletus) badius
(family Boletaceae) **Edible**

Description

The brown colouring of the cap closely resembles that of a chestnut. In the dry state it looks dull, but growing among damp moss it appears a glossy, viscid but pleasant reddish-brown. Since it is smaller and squatter than most boleti, this species cannot be mistaken for anything else. The tubes are a pale yellow, becoming greenish to pale olive. The whitening of the pores is a sign of old age. When bruised they turn a deep blue. The smooth, striped stem is a pale brown but is not veined as in the Penny Bun, nor does it become swollen when growing on firm, dry ground. At first the flesh is hard; with ageing it softens and collapses, beginning with the cap. It discolours only a delicate blue, if at all.

Habitat

This species inhabits woods of pine and fir, and only occurs in deciduous woods in which conifers happen to grow. In young woods which are often rich in boleti it does not appear, or only rarely. In forests which were planted not less than sixty years ago, it can be seen in masses. Quite often it grows at the base of living trunks or on decaying tree-stumps. Like the Penny Bun, it prefers acid soils, and its presence can be considered a reliable indication of acidity in soil. It always avoids chalk. The stoutest specimens usually occur under trees along the woodland border, and may be as large and stout as the Penny Bun. The true season begins when the nights begin to grow cool. Almost overnight the ground becomes covered with these dark-capped fungi, to the extent that one has to step carefully to avoid them.

Edibility

X. badius is one of the most sought-after edible species, although not as productive as the Penny Bun, alongside which it often grows.

Similar species

This species should cause no confusion with other boleti if one remembers the brown colour of the cap, the strong blue staining, when bruised, of the yellow-green tubes, and the slender build of the smooth, veinless stem.

PLATE 6

Xerocomus (= Boletus) chrysenteron
(family Boletaceae)   **Edible**

Description
In its early stages, this species is a deep brown, soon turning to a greyish olive. The edge of the cap has a pale fringe. The soft, felt-like surface does not turn greasy in wet weather; in dry weather, it splits open around the margin giving it a cracked appearance. The flesh which shows through slowly turns a cherry red. The same colour appears where slugs have been feeding. The tubes and pores are yellow, later turning a darker yellow with an olive tint. The stem is mostly a cherry-red, but can also become a clear yellow. In very young specimens, the stem may look swollen in the middle, or of an even thickness. It is finely streaked but not reticulated. The pale yellow flesh is juicy in fresh specimens so that small moisture bubbles appear on a cut surface. The flesh seldom turns blue, although the pores may do so if they are pressed.

Habitat
The widespread and even distribution of this species has no particular connection with any soil condition, nor with the trees it grows under. It may be found with oak, fir, larch or pine. It appears mostly in thick undergrowth, and is less common in open country. It can also be found along woodland borders and in parks of a woodland nature. It appears from June to November.

Edibility
Young specimens, like the smallest of those in the facing plate, are tasty to eat. Later the flesh becomes very soft, and is often full of whitish or golden-yellow patches of mould. The tubes should be removed immediately after gathering, as they quickly decay.

Similar species
A close relative, *X. subtomentosus* (*see* no. 156), can be mistaken for *X. chrysenteron*. The cap is olive, the angular tubes a clear golden-yellow and the stem is without the cherry-red colouring. Even closer to *X. chrysenteron* is its own sub-species, var. *versicolor*. This is a rare but beautiful purplish blood-red fungus which grows in grassy woodlands and on roadsides.

PLATE 7

Suillus (= Boletus) luteus
(family Boletaceae) **Edible**

Description

The cap of this easily identified boletus is a chestnut to chocolate brown in colour. It is of a distinct conical shape; and it becomes very slimy in wet weather. The yellow tubes come away easily from the thick cap. At first the tubes are covered by a veil which joins the cap margin with the stem. When the veil is broken it remains as a white ring, later changing to violet, near the stem apex. The pale butter-yellow flesh is tender and watery, and quickly decomposes.

Habitat

This boletus always grows in the neighbourhood of pine trees. A place where one would be likely to find it is under an isolated pine in a mountain meadow or sheep pasture. The mycelium will spread along the roots of the tree, and even beyond. It is actually joined to the root-hairs of the pine tree, a condition which occurs in many other species and which is known as mycorrhiza (*see* p. 26). The resulting fruit-bodies often appear in rows, or in a semi-circle (known as a fairy ring) as in the case of the Blewitts, *Rhodopaxillus* (= *Tricholoma*) *nudus* (*see* plate 60). *S. luteus* sometimes appears in masses on sandy or peaty soil. However, it does not avoid chalk.

Edibility

The flesh of this popular boletus, which looks so appetizing, has a pleasant, mild taste. Maggots attack it far less than they do the Penny Bun or *S. granulatus*. The slimy skin, which is easy to peel, should be removed before cooking. This species should not be left too long after gathering.

Similar species

Among other species with slimy brown or yellowish caps is the lovely golden-yellow ringed *S. grevillei* which is always found beneath larches. In colour, the chocolate-brown, unringed variety of *S. granulatus* (*see* plate 10) is not dissimilar, but has a pinkish base to its stem. The normal *granulatus* has a white stem as with *luteus*. A young specimen of *Gomphidius glutinosus* (*see* plate 14) is very similar in appearance, except that instead of tubes the radiating lines of gills can be seen through the thin cap.

PLATE 8

Suillus (= Boletus) aeruginascens
(family Boletaceae)

Edible

Description
The viscid cap can be pale grey, yellow-grey or pale brown, frequently tinted with green. The tubes and pores are pale grey, later becoming a dark grey-brown, and are angular in shape. The stem has a white ring which at first entirely covers the tubes, then disappears in old specimens. The white flesh turns a dirty grey-green when cut. If a piece of paper is wrapped around it, it will stain the paper a striking blue-green.

Habitat
This boletus grows exclusively among larches, rarely under other trees, unlike most woodland species which are associated with more than one kind of tree. Its habitat is on calcareous soil, and it avoids the more primitive rocks and sands. It has a mycorrhizal relationship with the larch. If it is not linked to the tree's rootlets, it cannot fruit. This is readily noticeable along woodland borders, wherever the larch appears. In open places where a field borders the woodland, and the ground is not covered with needles, the fungus will only spread to the limits of the tree roots. This is because the needles change the chemistry of the soil, making it acid. This appears to be necessary to some species of boleti. The mycelium grows around the root-lets forming a dense mat so that each root-ending is completely covered. This association appears to benefit both partners (*see* mycorrhiza, p. 26).

Edibility
S. aeruginascens is an edible fungus, although it is unpopular because of its softness and its viscid quality. A useful tip to remember, when picking toad-stools for the kitchen, is that all boletus species with rings are edible. This does not imply that all other species are inedible. However, it must be re-membered that it applies only to Boletaceae and not to other groups with rings where there may be a danger of poisoning—*e.g.*, *Amanita* toadstools.

Similar species
A related ringed species is the much rarer *S. grevillei* (*see* p. 46). Also to be found growing under larches is a species called *Boletinus cavipes*. It has a wavy brown, scaly cap and hollow stem. There is also *S. tridentinus* which has rusty reddish-orange tubes.

PLATE 9

Suillus (= Boletus) granulatus
(family Boletaceae)   **Edible**

Description
In the young specimen, the cap is curved over; gradually it spreads out, becoming wavy in shape, sometimes with an upturned edge. It does not normally exceed $4\frac{1}{2}$ in. in width. In damp weather or in dewy grass it becomes viscid, but in sunshine it soon turns dry and shiny. The greyish yellow colour then changes to an attractive golden-ochre. Beneath the delicate skin of the cap the flesh is soft and coloured white to yellow. At first the tubes are very short and narrow, with whitish yellow pores which may ooze a few milk-white droplets. Later the pore surface becomes a more olive-yellow. The tubes are easily removable. The fairly short stem has little granules at its apex. These turn brownish, and can then more easily be detected against the pale stem. When handled this fungus feels sticky.

Habitat
Like *S. luteus* (plate 8) this species is closely bound up with conifers. However, it can occur on the richer soils as well. Though it can still be found in May, it is looked upon more as a spring toadstool, growing among pines and junipers. Depending upon weather, it can remain in fruit until October, when the autumn frosts quickly kill it off. When growing in short meadow grass, this pale yellow boletus is often found in clumps, rows or semi-circles (*see also S. luteus* plate 8) wherever there are conifers. Whether it grows in long grass, in a young plantation, or is overshadowed by tall trees, it always appears close to conifers. It avoids clear sand or peat, and is confined to chalky and gravelly soils, particularly along woodland rides.

Edibility
As an edible species, this is worth looking out for. It is necessary to remove the sticky skin as soon as it is picked, and to cut the flesh through to see that it is free of maggots, which unfortunately is not often the case.

Similar species
S. granulatus may be distinguished from other slimy species by its flattened cap and ringed stem (*see* plate 8). There is a rare sub-species, var. *collinitus*, with a rose-coloured base to the stem.

PLATE 10

Suillus (= Boletus) variegatus
(family Boletaceae) **Edible**

Description

This is one of the commonest fungi, occurring in sandy, coniferous areas. The yellow, olive-tinted cap is covered with dark, sand-coloured granules. Before the granules appear, the cap has a very fine, felt-like surface. In damp weather it feels slimy. Young specimens are arched, and remain cushion-shaped for some time. Both the tubes and pores are an unusually dark olive with hardly any change in colour in ripe specimens. The pores are angular. The stem is cylindrical, swollen below, and somewhat paler than the cap. The yellow-to-ochre flesh turns a delicate blue when cut, and has a sour smell.

Habitat

This species is associated with pines and, like some other species—for example, *S. luteus* and *S. granulatus*—will not occur unless in their neighbourhood. It prefers acid soils and grows among the heather and heathberries where it is the most commonly found of the boleti. It even flourishes in beds of moss.

Edibility

This is one of the more valued of edible boleti. Although it is not as good to eat as the Penny Bun, it can be used as flavouring in many dishes.

Similar species

The granulated skin, the dark olive tubes, and the fact that it is to be found growing among pines, should make identification simple. Another species often seen in the same place is *S. bovinus*. Its small, clear, yellow-brown cap is sticky, very smooth and rubbery—a condition rarely found in other species. The olive tubes are strikingly white and angular. It grows in large groups. An interesting feature is the frequent association with a species *Gomphidius roseus*, an agaric about $1\frac{1}{4}$ in. across with rose-red gills. It is a relative of *G. glutinosus* (plate 14). *G. roseus* grows mostly with *S. bovinus*, and is so well-hidden that one often only notices it when it is kicked over. Should it be found growing by itself, then one may be sure that *S. bovinus* is near at hand. Mycologists have not yet discovered the reason for this close association.

PLATE 11

Paxillus involutus
(family Paxillaceae)   ☺ **Edible if well cooked**

Description

The cap, which may reach a width of $4\frac{1}{2}$ in. or more, is yellow-brown with olive tones. The raised centre later hollows out. The rim is strongly curved over, and remains so throughout. This gives the young specimens a rather stately appearance. When moist the skin is sticky and can be peeled back from the rim for some way. It has a felt-like texture at the rim. On the under-side of the cap are the fine and crowded gills; but in spite of the fact that this is a gilled species, it is more closely related to the genus *Boletus* than to the agarics. In time the gills become clogged and stain a noticeable dark brown. The gills run down the stem (decurrent) where they join up loosely in a reticulated, pore-like network. The pale brown, often reddish stem forms dark brown patches where it is injured. The flesh of the cap is pale yellow and juicy, somewhat reddish in colour in the stem.

Habitat

This toadstool grows in the same kinds of conditions as *Xerocomus badius* (plate 6). It enjoys damp, even swampy, ground under pines and firs, and is found in bogland, but does not appear in deciduous woods or on chalky soils. It fruits mainly in the autumn months.

Edibility

This is a well flavoured edible species, but it must be treated with some caution. Raw and undercooked specimens can produce gastric upsets. Well cooked, the tender flesh has an exceptional taste, even though, when raw, it may have a dull and unappetizing appearance due to bruising.

Similar species

There is a pleasing yellow sub-species *leptopus* which has a distinctive scaly cap and clear yellow flesh. It grows close to alder, and is as edible as *Xerocomus badius*. Growing among pine woods is another species, similar in appearance: *P. atrotomentosus* (plate 13), which is not much valued.

PLATE 12

Paxillus atrotomentosus
(family Paxillaceae)   **Inedible**

Description
The attractive brown cap, which later becomes paler, is at first cushion-shaped, then flattens out and can reach a diameter of 6 in. The rim remains curved over for a long while. The yellow gills become brownish with age. The gills are decurrent and veined near the stem, forming a flat surface of irregular pores, and resembling those of a boletus. This species also resembles a boletus in that the gills separate more easily from the cap than in other agarics. This pored feature and the looseness of the gills would suggest that there is an affinity between the two families Agaricaceae and Boletaceae. A special feature of this toadstool is the dark, brownish, felt-like covering on the bulbous stem, which is firm and squat in shape. The white flesh looks appetizing but tastes disagreeably bitter, and is very watery.

Habitat
This toadstool is confined to conifer woods where it lives on pine and fir stumps, and is one of the commonest of the woodland species. The irregularly shaped fruit-bodies appear quite suddenly above the ground on rotting wood, from July onwards. In dry summers, it requires very little moisture, unlike most toadstools which can hardly survive without it.

Edibility
It is a pity that this easily recognized toadstool, with its attractive maggot-free flesh, is far from being tasty. By boiling it and pouring off the liquid it can be made more edible. However, because of its indigestibility it is not to be recommended. A species with a far better flavour is *P. involutus* (plate 12). This, too, frequents coniferous woods, where it grows on the bare soil. Only occasionally is it found on tree stumps.

PLATE 13

Gomphidius glutinosus
(family Gomphidiaceae)   **Edible**

Description
The brown, slate-grey or violet coloured cap of this easily recognized toadstool is covered with a thick, colourless, transparent, slimy veil. In young specimens the veil stretches from the border of the cap to the stem, covering the gills like a window. This tears away to form a ring-like pad round the stem. The spores are black, and they turn the white gills grey and finally black. The gills are thick and waxy, moist and decurrent. The stem is white above and pale chrome-yellow below. The slimy ring on the stem also turns black from the falling spores. The flesh remains a pure white except in the lower stem where it is a beautiful chrome-yellow.

Habitat
This characteristic species commonly occurs in pine woods, from summer to late autumn. It grows in scatterings and not in groups and is especially fond of plantations with a good carpet of needles. It is seldom seen among other conifers, and never in deciduous woods. It can grow in any kind of soil of a turfy nature.

Edibility
In spite of its unappetizing sliminess, this is a very good eater when mixed with other toadstools. In spite of the fact that it is soft, watery and not very strong tasting, it deserves to be used more. As soon as it is picked the slimy skin should be removed with a knife. It can be peeled off the cap in one movement.

Similar species
It can hardly be mistaken for any other toadstool. Seen from above, it might resemble *Suillus luteus*; but the gills and stem will distinguish it (*see* plate 8). There is a short variety with a pale grey cap and no sliminess, and with paler gills. Sometimes young specimens are coloured a rosy red. In this small genus the next commonest species is *G. rutilus*. It has a reddish brown stem and is found mainly under pines.

PLATE 14

Hygrophorus chrysaspis
(family Hygrophoraceae)   **Edible**

Description

This is a strong-smelling species. The cap is at first a clear white, then becomes a chrome-yellow mainly along the margin, as occurs with the gills. When dried out it turns almost orange. The irregularly shaped cap has a slimy, dirty look. The apex of the similarly coloured stem is covered with tiny white pustules. In wet weather it is so slippery that it can hardly be held between the fingers. The white flesh emits a characteristic odour similar to that made by the goat-moth caterpillar. Once this larva has been smelt the resemblance is very striking. This species has a wide variety of forms. In addition there is a pure white one, with no special odour, which can be treated either as a separate species or merely as a variety of the coloured kind.

Habitat

As early as the summer, this remarkable toadstool can be seen in large clumps both in deciduous and conifer woods. It avoids bare and acid soil.

Edibility

This species is quite edible, although seldom gathered because of its smell and slimy nature.

Similar species

It is highly unlikely that this species could be mistaken for any other. Similar looking species have thinner gills, and are neither slimy nor sticky. Care should be taken, however, not to confuse it with another white species —namely *Clitocybe dealbata* (plate 51). There is also a relative, *H. penarius* which has a sturdy stem, no slimy cap, and a weak smell. Another species *H. chrysodon* has yellow-gold tubercles on the border of its cap and on the stem apex. The species *H. piceae*, when still young, is distinguished by its strong white cap, delicate yellow gills, and a soft and agreeable smell. It is to be found in pine woods in the mountains. Finally there is a species *H. cossus* which has a soft yellow centre to its cap, but, unlike *H. chrysaspis*, does not change colour. It has the same smell as the goat-moth caterpillar.

PLATE 15

Hygrophorus hypothejus
(family Hygrophoraceae)   **Edible**

Description
Against a carpet of pine needles, the olive-brown caps of this late-fruiting species are hardly noticeable, yet there may be dozens of these attractive toadstools growing together in the pine wood. The dark, yellow-tinted cap is slimy, as is the much paler stem. In young specimens the turned-over margin of the cap reaches the stem. As it flattens out, becoming more funnel-shaped, a soft, pale trace of a ring is left behind. The white gills take on a yellow-orange sheen. The more the cap spreads out, the lower down the stem the gills run (decurrent). Exposed flesh slowly turns a reddish yellow. There is only a slight odour.

Habitat
This species appears in groups in autumn, but only when the night frosts begin, and can be seen up till December, and even later. This species is found under pine trees, and rarely elsewhere.

Edibility
This is a good edible toadstool, in spite of its slimy nature. When most toadstools have died away *H. hypothejus* can still be harvested, and will provide a tasty meal.

Similar species
Also among pines can be found *H. olivaceoalbus* whose cap and stem is coloured a strong olive. The stem is also banded or speckled. The gills remain white. Apart from its typically slim form there is also a much sturdier variety, *obesus,* which could be taken for a different species. It grows under pine on chalk, has a similar olive colouring but a much fatter, white stem. It is edible and much more common. The yellow form of *H. hypothejus* can be mistaken for another species *H. lucorum*. This is to be found growing with larches and nowhere else, and is always a clear yellow. The stem is much paler and the gills have an odd lemon-yellow lustre. A further species, *H. agathosmus,* has a pronounced grey cap and smells of almonds. Tiny warts may occur at the centre. This species is very common and widespread in mossy places in coniferous woods, where it often forms fairy rings. A very rare variety *aureofloccosus* has golden-yellow marks at the stem apex, and an especially beautiful appearance.

PLATE 16

Hygrophorus conicus
(family Hygrophoraceae)   **Edible**

Description

This species has a striking, fiery-red cap with a pronounced central boss. Although of only average size—about 3 in. across—its bright colour quickly attracts the eye. It is distinguished from similar species by the gradual blackening of grown specimens (and of young ones after they have been picked). At first there are only streaks on the cap, on parts of the gills and the stem, and these gradually darken. The thick, soft gills are set wide apart and are conspicuous in their yellow colouring. The stem is at first a pale yellow-red. Later the blackened toadstool stands out beside its young and as yet untouched neighbours. The flesh of this genus is rich in juice, and is fragile and transparent.

Habitat

This species is not found in large numbers. It occurs mostly in rainy summers and autumns. Like other, similar species it is not found in closed-in woodland, but in the grassy borders of woodland paths, in glades or orchards, or beneath scattered trees in parks. It is less frequently to be seen in mown meadows.

Edibility

All *Hygrophorus* species give a mild-tasting flavour to soups and vegetable dishes. They are only worth picking when found in luxuriant vegetation. These toadstools are easily crushed and care should be taken when placing them in the basket with other specimens.

Similar species

Of the numerous species in this genus, the largest is *H. puniceus*. The handsome red, cone-shaped cap may reach a width of 6 in. All members of the sub-genus *Hygrocybe* have smooth, or at most fibrous, stems which are streaky, not spotted, at the apex. Stems are slender, almost reed-like. Many species, such as the small, entirely green *Hygrophorus psittacinus*, are very slippery. These toadstools are often found growing in short grass on hillslopes, and this type of place can be a rich hunting-ground. Another sub-genus with many species is *Camarophyllus*, with predominant white, grey, ochre and brown colours.

PLATE 17

Lactarius deliciosus var. *semisanguifluus* Saffron Milkcap
family Russulaceae) **Edible**

Description

This is a very easily recognized toadstool because of its carrot-red colour-ing which first appears most strongly on the cap and gills. The hollow stem may come as a surprise when cut (as shown in the illustration). The greasy cap is frequently marked with dark orange or purple-grey rings. Green markings which appear are a distinguishing feature of the true Saffron Milkcap. Injury caused by handling, or the effects of autumn frosts, will increase these green markings. Often young specimens are so heavily marked with green, that only by breaking the stems are they recognized. The gills are extremely brittle, a feature of this family (Russulaceae). The short stem is often marked with green and is always hollow. The flesh is soft, especially in specimens which have lost milk due to cutting. Taste is mild—which is unusual, since members of this genus usually have a sharp or bitter flavour.

Habitat

The Saffron Milkcap occurs with conifers and is not found in purely deciduous woods. Often an astonishing number of these toadstools can be discovered in young hill-plantations. They will continue to grow there until the trees are finally cut down. Where this toadstool is found growing in a deciduous wood, you may be sure that there is a conifer growing near by, which may be hidden by other trees, and so overlooked. Along woodland borders the toadstools grow out along the tree-roots, a feature also noticed in the boletus *Suillus aeruginascens* (plate 9). It prefers either sandy or chalky soil. Some specimens may turn up as early as June, but the main season is the autumn. If the summer crop is badly attacked by maggots, then the best gathering would be in late autumn.

Edibility

This species, which is widely collected and much treasured, is best fried or baked in fat or oil. It is not recommended for soups and vegetable dishes, in which it has an unpleasant and rather penetrating flavour.

Similar species

L. deliciosus is very similar to *L. sanguifluus*, but the latter can be distin-guished by its dark, wine-red milk, firmer flesh, and its habitat—it grows in chalk, under conifers. It is also edible. Other species which have white and sharp-tasting milk can be edible after frequent washing. A further variety is var. *salmonicolor*. It shows no green colouring, grows to a far larger size, and appears later in the year.

PLATE 18

Lactarius torminosus Woolly Milkcap
   (family Russulaceae)   ☠ **Inedible**

Description
The "woolly" covering on the pale or flesh-red cap makes this species easy to identify. The curved border frequently carries a ragged fringe of skin. When the colour-zones on the cap are clearly formed, this toadstool can look very much like the last (plate 18). However, the difference lies in the fact that its gills are white, not saffron coloured. The stem is similar in colour to the cap, although somewhat paler, and is hollow. Dark, rounded patches may appear on it. The white, or dull red flesh, releases a white milk which has a sharp and burning taste. The gills remain active longest in dry weather.

Habitat
This is a birch wood species; it has a mycorrhizal link with the birch tree and does not occur under other trees. If found under beech, then there is always a birch tree or two in the vicinity. It does not turn up in pure beech or pine woods.

Edibility
This is not recommended as an edible toadstool. It has been called poisonous because of some cases of gastric upset, but it can, in fact, be used by first washing and cooking, so as to remove the milk. Even so, it is not recommended.

Similar species
The resemblance to the real Saffron Milkcap is very close, but the carrot-red milk of the latter helps to distinguish it. The species *L. zonarius*, which has white milk, might also be mistaken for it, but here the cap is bare and slimy. The colour is straw-yellow with concentric rings. It grows in deciduous woods and is not uncommon on soft, chalky soil. *L. torminosus* has a subspecies of a paler colour, but with the same kind of ragged cap. This is var. *pubescens* which is smaller, whitish to ochre in colour, and found on acid and peaty soils. It also inhabits birch woods. It is best not eaten.

PLATE 19

Lactarius volemus
(family Russulaceae)   Edible

Description
This is a fairly common species but is often misidentified. One feature to note is the rich supply of white milk which oozes out of any cut, and slowly turns brown with exposure. The reddish or yellow-brown cap can grow to a considerable size. At first curved over, it later deepens into a more irregular shape. The dry skin is at first downy, then smooth. The pale yellow gills become browner with age. The flesh also browns where it is eaten by slugs, as can be seen in the left-hand specimen in the illustration. The stem of this species is much sturdier than that of other near relatives. The flesh has a curious smell resembling that of herring, and gives off an unpleasant odour. The smell will cling to the fingers, if picked, and it can ruin the flavour of other toadstools. The taste is mild but unpleasant.

Habitat
This species does not require any special soil or mycorrhizal association, and occurs as frequently among conifers as under beech trees. It is often found in company with the much larger and strikingly white *L. piperatus* (plate 22), with *Russula cyanoxantha* (plate 24) and with the pale deciduous woodland form of the Chantarelle, *Cantharellus cibarius* var. *pallidus* (plate 63), as well as the Horn-of-Plenty *Cratarellus cornucopioides* (plate 66). In many regions, where *L. volemus* is rarely found, other species may be mistaken for it. They probably look slimmer in build, however, and will not contain the brown milk.

Edibility
This fungus is ideal for frying, like all the orange milkcaps, whose sharp or bitter taste can be removed by cooking.

Similar species
Among the more valued brown species is *L. rufus* (no. 136), found in sandy and turfy woods, often in large numbers. The cap is a reddish-brown and has a central boss. The milk is sharp and burning but can be removed by washing so as to make the flesh appetizing. Very similar in appearance, but less appetizing, is the smaller *L. mitissimus* (plate 21).

PLATE 20

Lactarius mitissimus
(family Russulaceae) **Edible**

Description
Among the numerous species of smaller milkcaps which have a brownish appearance, this species is outstanding because of its orange colouring. It may sometimes be mistaken for a young *L. volemus* (plate 20). The cap is somewhat sticky and is evenly coloured throughout. It soon becomes flattened and can reach a width of $2\frac{1}{2}$ in. In old specimens the colour fades. The gills are adnate, with a small hook next to the stem. They are thick and crowded, coloured a fleshy yellow which darkens with age. When ripe, the spores give the gills a powdery look. The thin, delicate, clear orange stem is soft and breaks easily. The pale red flesh gives off a white milk when cut. The taste is not sharp as in other milkcaps, but rather bitter and irritating to the lips.

Habitat
This species first appears in groups at the beginning of autumn, commonly under fir trees, where the caps stand out against the dark needles or bright green mosses. It may also occur in pine woods. As it pales with age to a dull orange, it may become more difficult to identify. It fruits up till the end of November and is one of the most long-lasting of the milkcaps.

Edibility
The edible quality of this milkcap is poor. Its small size makes it hardly worth collecting. If used, it should be steeped in water for some time, so as to remove the resinous and bitter part. It should be preserved in salt or vinegar. This applies to most milkcaps, apart from the genuine Saffron Milkcap (plate 18).

Similar species
Apart from the much sturdier *L. volemus* (plate 20) there is a whole range of edible milkcaps. Similarly coloured is *L. porninsis* which has a large, zoned cap and a sharp-tasting milk. It is found only under larches, and is rather uncommon.

PLATE 21

Lactarius piperatus
(family Russulaceae)   **Edible if prepared with care**

Description
This toadstool can be recognized by its whitish colouring and the size of its cap which is often considerable, reaching 8 in. or more. The cap is smooth or slightly wrinkled. It looks dull in dry weather but becomes slimy and shiny after rain. The deep hollow centre turns an ochre colour and the skin breaks up into irregular patches. Specimens growing in damp situations, however, remain white and do not tear. The gills are very crowded with hardly any space between, especially in young specimens. They vary in length between stem and cap border, and are a clear white, turning a dirty yellow with age. The white stem is slightly wrinkled, and is of variable but often considerable length. When squeezed, the firm flesh oozes drops of white milk which has a sharp and burning taste. The drops dry slowly; and in a variety called *glaucescens*, they turn a grey-green.

Habitat
L. piperatus inhabits the same type of soil and woodland as *Russula cyano-xantha* (plate 24). These two species are commonly found together between June and September. Many kinds of brightly coloured toadstools, including many of the family Russulaceae, are found in beech and oak woods where there is a scattering of hornbeam and copper-beech. In these woods may also be found the Wood Hedgehog *Hydnum repandum* (plate 69), the *R. emetica* (plate 27), the mild tasting *R. rosea*, the *Clavaria pistillaris*, and the common, dark grey Horn-of-Plenty *Cratarellus cornucopioides* (plate 66).

Edibility
Although not poisonous, this species requires much careful preparation before it is appetizing. Baking, or cooking with other ingredients in pastry, will improve the flavour.

Similar species
The very similar *L. vellereus* (no. 143) is quite uneatable. Its dirty white cap on a short, fat stem, and wide-spaced gills will identify it. Of similar colour and build is *R. delica* which has a whitish cap and white, mild-tasting flesh but without any milk. The blue-green sheen on the gills, and their contact with the stem (adnate), helps in identification. This is an edible species but with a poor taste.

PLATE 22

Russula vesca
(family Russulaceae) **Edible**

Description
The main feature of this toadstool is the pallid to deep flesh-red colour of the dark veined cap. This colour later disappears from the rim of the cap, revealing the flesh. (This can be clearly seen in the large specimen in the illustration.) The gills retain their whiteness but become flecked with rusty spots. The typical russula-like brittleness of the gills which snap when broken is not very noticeable in this species. The firm, white stem has a pointed base which is buried in the soil. The flesh is white, soft when dry, and has a mild and agreeably nutty taste.

Habitat
This species is not connected with any particular woodland tree, and is found both in deciduous and coniferous woods, preferring a sandy soil. It is particularly common where heaths and bilberries grow, also between lichens and below bracken. This is also the habitat for the larger, rusty-brown *R. adusta* (no. 124), the Blusher *Amanita rubescens* (plate 34), and *Rozites caperata* (plate 47).

Edibility
Of the many much-prized russulas, this species is the favourite. Unfortunately it is often attacked by maggots. It is best to remove the slightly sticky skin of the cap as soon as possible after gathering. It will peel off from the margin inwards.

Similar species
It is possible to confuse this species with close relatives, unless all the characters are closely observed. However, there is no risk of poisoning from similar species. The fear which exists, that this toadstool is poisonous, is groundless.

The genus Russula is among the commonest and most outstanding of woodland toadstools. During the summer they appear in large numbers and varieties. From a distance, their caps of red, yellow, violet or green stand out clearly. However, colour can be an unreliable guide to identification. More typical is the soft, dry and cheesy quality of the flesh which is without milk. The brittle, easily broken gills are white, often becoming a dark yellow. Most of the eighty or so species are edible. Sampling the tiniest piece of gill will tell in seconds whether the toadstool is edible or not. A mild, nutty taste means that it is eatable; a sharp, burning taste signifies that it is not. Two of these bitter species are shown on plates 26 and 27.

PLATE 23

Russula cyanoxantha
(family Russulaceae)   **Edible**

Description

The cap, which in a mature specimen will be 6 in. or more in width, becomes either lilac or green, or a mixture of both. On this are superimposed spots of ochre or pale yellow. The young toadstool has a ball-like shape, then flattens out, to become funnel shaped. The skin of the cap is damp and shiny, and peels away from the edge. The white or lilac coloured flesh can be seen underneath. The pure white gills are exceptionally soft in this species, and do not snap off in the manner of other russulas. The stiff, ringless stem and the soft, dry flesh leave no doubt as to the genus of this toadstool. The taste of the flesh is mild, like that of *R. vesca* (plate 23) but if a piece is chewed for some time it develops an alkaline taste. The club-shaped or unevenly cylindrical stem is weakly furrowed. It is usually white, but in rare cases is lilac-tinted.

Habitat

This species is found mainly in beech woods or in other deciduous woods. It appears in June, often in groups but also spread widely over its habitat. Even a single beech tree will attract it. It prefers chalky soil and is seldom seen in coniferous woods. It may fail to appear over wide areas where its place is taken by similar species.

Edibility

This is an edible russula. Its flesh is not as dry as that of other species and its large size makes it worth searching for. Like all mild-tasting species it is much attacked by maggots and slugs. Because of its size it should be cut in half before it is placed in the basket.

Similar species

Characters which distinguish it from other species are the ringless, soft, crumbling and furrowed stem, the swollen base, and the lilac tinge to the cap. Care should be taken not to confuse the pure green form with the most deadly of toadstools—the Deathcap *Amanita phalloides* (plate 29).

PLATE 24

Russula integra
(family Russulaceae)   **Edible**

Description

This is one of the commonest of russula species found on chalk. The colour of the cap can be brown or brownish purple, or purple-red, or sometimes violet. The colour changes towards the centre of the cap into a bright yellow, or into a paler patch. Colour is not definitive in this toadstool. The skin is wet and sticky, with a shiny look even when dry. It appears commonly among pine needles. The rounded cap of the young specimens are half-hidden in the needles, so that as the cap spreads out into a funnel shape, needles and fragments of humus get carried up with it. The edge of the cap is blunt, then becomes furrowed. Gills are at first white, soon taking on a pale yellow colour, finally becoming an ochre to brownish-yellow. They are widely spread on the outside, and are very brittle. The stem is compact, white all over, somewhat wrinkled, with a firm skin which later becomes soft. The flesh has no particular smell.

Habitat

This species appears in groups as early as July, and continues to fruit until October. It prefers coniferous woods where it is one of the commonest species, growing on a carpet of needles in chalky soils. It is as much at home in old woods, with a carpet of moss, as in young plantations. Two very similar looking species, which grow in deciduous rather than coniferous woods, are *R. alutacea* and the large *R. olivacea*.

Edibility

This well-distributed species can be considered a good edible fungus, although its hard and poorly flavoured flesh does not appeal to all tastes.

Similar species

Since other russulas can be mistaken for this species, the yellow gills should be taken into account. With *R. firmula* the flesh should be tasted. It is edible in spite of its rough, sharp flavour. The cap rarely grows beyond a width of 4 in. and the colour is more of a deep violet. However, it is best to learn the slight differences between *R. integra* and *R. firmula* by making a visual comparison, since they both taste the same. A similar, yellow-gilled species *R. badia* has a more purple tinge to its dull cap, a somewhat reddish stem, and emits a smell of boxwood. The flesh has a sharp, burning taste. It is inedible. There are many other similar species and their classification requires a detailed study.

PLATE 25

Russula queletti
(family Russulaceae)   💀 **Inedible**

Description
The reddish-lilac colour of the cap is strongly spotted. Often olive tints appear. The colour is useful for recognition if all its variations are borne in mind. At first the cap is rounded, then flattens out and becomes hollow in the centre. The rim is wavy and striated and the skin is sticky and easily removable. Its colouring and the brittleness of the gills help to distinguish this russula from close relatives. The gills remain white for some time, and only in adult specimens turn a clear butter-yellow. The stem is a purplish wine colour, and has a weak/white ring which comes off when touched. In wet weather, young specimens give off water droplets from the edges of the gills. The soft, cheesy flesh has a sharp and burning taste. Its fine, fruity smell has been likened to that of gooseberries.

Habitat
R. queletti is closely associated with pines, often in company with the Saffron Milkcap. It also occurs frequently under conifer hedges and with single pines in parks. It fruits from August to October on chalky and sandy ground, but avoids peat.

Edibility
This is not an edible species. *R. queletti* and its closest allies have sometimes been classed amongst the poisonous species.

Similar species
Clear, olive coloured, soft or firm-bodied examples can be mistaken for other species: for *R. sardonia*, for instance. However, this latter species grows with conifers on acid soils. Its very watery gills are at first a clear lemon-yellow, and slightly decurrent. A slight difference in smell is soon detected. The whole of this russula is firm-fleshed and lasting. It is also inedible. A clear, rose-red cap and stem is found in *R. sanguinea* (no. 132), also a common pinewood species with a sharp taste. The pale yellow gills are clearly decurrent and turn a strong lemon-yellow when damaged. One of the best of the mild tasting edible species is *R. xerampelina* var. *erythropoda*. It has a deep blood-red, finely felted cap, a reddish wrinkled stem and yellow gills. The flesh slowly turns brown when cut and gives off an unmistakable smell of herring.

PLATE 26

Russula emetica Sickener
(family Russulaceae)

☠ **Inedible**

Description

This is one of a whole series of red or reddish coloured, sharp tasting russulas. The real Sickener has a clear blood-red, often yellowish or white cap, and young specimens can be quite colourless. The colour also sometimes fades after rain. The sticky skin can be peeled back from the raised and furrowed rim. White gills in many specimens take on a bluish or yellowish lustre. The spore dust which falls is pure white. The stem is white, sometimes a faint red. The skin is often wrinkled. Sometimes the white flesh is full of holes and, like that of the cap, very spongy. It has a peculiar, fruity smell, and a sharp, peppery taste which is felt immediately on the tongue.

Habitat

The Sickener appears in summer in beechwoods, growing in the leaf litter as well as along mossy pathways and on beech stumps. On marshy soil in coniferous woods a bell-shaped form, with a lighter red cap, occurs.

Edibility

The bitter taste of this species can be removed with cooking, but this is not really an edible toadstool, and is unsuitable for sensitive palates. In some of the early books on the subject it is named as highly poisonous, and listed among the most dangerous species.

Similar species

The smaller *R. fragilis* (*see* no. 134) which occurs in many shades of red, often with a clear olive-green centre, grows in groups, especially in coniferous woods. It is also inedible. Another species, *Russula rosea*, occurs in beechwoods. It is no different in colour, but duller and drier in texture, and not so shiny. The white stem is club-shaped and, seen under the lens, finely grained. It has a mild flavour but is not particularly tasty. Another red species is *R. luteotacta*, very similar in shape and size to the Sickener. The gills are wider apart. They eventually show, as on the stem, a number of clear chrome-yellow spots. The habitat of this species is near oak. All the russulas can be identified by their soft flesh which is never fibrous, but is more like cheese. Because of their wide distribution and bright colours they are easily found. Anyone who is familiar with this large genus, and who wishes to collect edible specimens for cooking, should either check the taste with a small piece from the gills, or check on the specific characters which will identify the species.

PLATE 27

Coprinus comatus Shaggy Inkcap or Lawyer's Wig
(family Coprinaceae) **Edible**

Description
The cap of this striking toadstool is always taller than it is broad. A total height of 12 in. is not unusual. At first, little of the stalk can be seen as it is entirely enclosed in the club-shaped cap. The outer skin of the smooth crown soon breaks into rough white or brownish scales. A rosy tint appears at the base of the crown, spreading upwards, and soon darkening. Finally, a complete liquidization of the cap takes place. The white gills remain very crowded and change colour, the colour rising upwards, through rose to black. With each falling black droplet thousands of black and minute spores are released. The white stem carries a small, temporary ring where the cap has separated, and is hollow inside. The white flesh is fibrous and has no particular smell.

Habitat
This species is an inhabitant of made-up ground, rubbish dumps, old compost heaps, as well as freshly sown patches of grass. New parkland and sports grounds can produce hundreds of these toadstools almost overnight. In woodland it is confined to the sides of pathways. It can often be seen on grassy banks. This species does not appear to thrive directly on dung heaps, which is the normal habitat of the smaller species. It appears from May onwards.

Edibility
In the young stage, when the gills are closed and the cap is still white, this species is very tender and appetizing. However, it must be used immediately after gathering, as it quickly blackens and melts away. Its remarkable growth-rate makes it possible to gather a crop every two days. At one time, the black liquid was used as writing ink.

Similar species
Other white species are much smaller, and of little use for the collection. A silver to ash-grey species is *C. atramentarius* (*see* no. 92), which occurs mostly in thick clusters along roadside banks, on grassy plots and waste ground. It is not as large as *C. comatus*, but melts away in the same ink-cap fashion, through a process of self-digestion, called autolysis. *C. atramentarius* can also be eaten when young, but alcohol should not be taken with it as in some people this can cause a peculiar reddening of the skin.

PLATE 28

Amanita phalloides Deathcap
(family Agaricaceae) ☠ **Deadly poisonous**

Description

The cap of this most poisonous of all toadstools develops half below ground inside an egg-shaped sheath, the volva, wrapped around the swollen base of the stem. Portions of this white skin remain attached to the sticky cap. At first the cap is entirely white, but soon turns an olive-green with fine, dark, radial streaks. It can grow to 3–5 in. in width. Pure white forms occur now and then. Gills remain white, at most turning a yellowish green. They do not reach the stem (free) and the edges are finely haired. The slender upright stem rises from its conspicuous, bulb-like base, which remains half or wholly buried and can easily be overlooked. The colour of the stem varies from white to a strong olive-green. In most cases the skin on the stem breaks into a characteristic banded pattern. A membraneous partial veil at first covers the gills, then breaks away from the cap-border to form a ring, on which grooves made by the gills can be seen. The flesh is pure white, with just a faint green tinge beneath the skin of the cap. It has a sweet, heavy smell of honey.

Habitat

The Deathcap appears in August, at first singly or in small groups, and continues to fruit until October, mostly near oaks and more rarely under beech. A light humus on sandy or limy soil is preferred. Often it grows in company with the edible species *Agaricus silvicola* (plate 38). Beware not to confuse the two! The Deathcap must never be eaten.

Edibility

The poisonous effects of the Deathcap are only felt some time after it is eaten, so that help may come too late. The poison is haemolytic and destroys the blood-cells, leading in most cases to death (*see also* p. 24). A single toadstool can kill a whole family.

Similar species

The False Deathcap *Amanita citrina* (plate 31) is often mistaken for the true Deathcap. It smells of raw new potato, and this was once considered to be a safe distinction when identifying these white amanitas. A far safer method is to look for the white gills and ring on the stem. Very close to *A. phalloides* is the pure white Destroying Angel *A. virosa* (plate 30). Remember that the Wood Mushroom *Agaricus silvicola* mentioned above has rose-coloured to violet-brown gills. The species *Tricholoma flavovirens* (plate 58), and young puff-balls, look similar, also, and can be confused with the dangerous Deathcap.

PLATE 29

Amanita virosa Destroying Angel
(family Agaricaceae)

💀 **Deadly poisonous**

Description
Like the last species, this club-shaped amanita is among the most deadly of toadstools. The sticky cap is white and egg-shaped in the young specimens, and emerges on a slender stalk. The high crown later turns more yellowish, and breaks through the sheath, carrying up pieces of skin. The half-buried, swollen base is surrounded by a volva with a ragged edge. Gills remain pure white throughout. They are free of the stem and may carry patches of the partial veil on their edges. The ring on the stem is usually damaged and often disappears early. The stem surface is flaky and much more ragged than the banded marking on the Deathcap. The pure white flesh has a disagreeably putrid smell.

Habitat
This species is far less widespread than the Deathcap. Coniferous areas seem to be its favourite habitat. It looks very conspicuous among the beds of moss or on the bare needles. It appears from early summer onwards in small groups, or singly, on sandy or gravelly soil.

Edibility
This species is just as dangerous as the Deathcap. Because of its fairly widespread occurrence and deadly qualities it is important to be able to recognize it.

Similar species
The closest relative is *A. verna*, the Fool's Mushroom (*see* no. 69). It is found on chalk but not in numbers. The arched cap, which later flattens out, is small, and the tattered looking volva is much firmer. This brings it close in appearance to the white form of the Deathcap. Young specimens closely resemble those of the Wood Mushroom, *Agaricus silvicola* (plate 38). The colour and shape of the caps look the same, but in *A. silvicola* the gills are rose to violet-brown instead of pure white. The edible *Amanitopsis vaginata* (no. 76) also looks similar in the white form, but here the edge of the cap is striated, and it loses its ring. It would be wise not to collect any of these amanitas at the same time as you are collecting edible species. Incidentally, when gathering puff-balls which can resemble young amanitas, a cross-section will soon reveal that they contain flesh, not gills.

PLATE 30

Amanita citrina (=mappa) False Deathcap
(family Agaricaceae)   ⊛ **Mildly poisonous**

Description

The cap of this widespread amanita is normally a yellow-green or pale lemon, but it can also be whitish. The young toadstool is entirely covered by the universal veil, but this soon ruptures so that the cap carries up with it a number of torn pieces as brown map-like patches (hence the former name of *A. mappa*). These sometimes become rubbed off. The white gills do not quite reach the stem (free). The slender stem is white, smooth and fibrous, and acquires a distinctive zoning or speckling. A temporary partial veil covers the gills at first, then becomes detached from the margin of the cap, and remains as a ring on the stem. The stem-base is swollen and mostly buried. The white flesh has a strong odour of new potato. Young specimens which happen to develop with crooked stems soon adjust to an upright position so that the cap remains parallel to the ground. In this way, the gills can liberate their spores without hindrance.

Habitat

From late summer onwards the fruit-bodies appear in large numbers among the heather and heath-berry plants, under pine, fir or birch trees; less frequently under beech. Acid soils are preferred, especially in sandy areas, in company with *Xerocomus badius* (plate 6), *Rozites caperata* (plate 47) and the Fly Agaric *A. muscaria* (plate 32).

Edibility

Once thought to be very dangerous, the False Deathcap has turned out to be fairly harmless, and at most only mildly poisonous. Its nutriment value is negligible.

Similar species

The very commonplace *A. gemmata* differs in having a grooved rim to its cap, white spots and a weakly swollen base to its stem. The ring is frequently absent. It occurs from June onwards on sandy soil near conifers. Although harmless it is not recommended for eating. At one time no distinction was made between the Deathcap and False Deathcap, on the grounds that both gave off the strong potato-like smell. Actually, the Deathcap has more of a repulsive honey-like odour. The loose volva, and sparse patches on the olive-green cap of the Deathcap are further distinctions. Pure white forms of both these species could be confused with the pure white *Agaricus* species (mushrooms). However, these have reddish-grey to coffee-brown gills, and there is no volva and no patches on the cap.

PLATE 31

Amanita muscaria Fly Agaric
(family Agaricaceae)
☠ **Poisonous**

Description

With its vivid red, white-spotted cap, the Fly Agaric is one of the most striking of all woodland toadstools. It starts as a white, wart-covered button, and at first nothing is visible of the red colouring (see the baby specimen in the illustration). As it develops, the universal veil splits and is conveyed onto the top of the cap which it covers in regular rows of white wart-like patches. These may wash off during rain. Under the skin of the cap may be found a strong lemon-gold layer. In its final stage this toadstool turns from fiery-red to a yellow. The white gills are broad and crowded, and do not reach the stem (free). The membraneous ring is heavily scalloped, and the stem looks white and woolly. Around the swollen base are more rings. The pure white flesh has no characteristic smell or taste.

Habitat

Pines, firs and birches are the commonest trees beneath which the Fly Agaric appears. It occurs in groups, in rows, or even in circles, growing in between moss-covered sand or turf, or in the bare needle-carpet. It is widespread, and is a reliable indicator of the presence of the much sought-after Penny Bun *Boletus edulis* (*see* plate 1). Half-grown conifer plantations, mixed with birch, is the ideal habitat for the Fly Agaric, and here, also, *Suillus piperatus* and the Chantarelle *Cantharellus cibarius* (plate 62) may be found.

Edibility

The poisonous properties of this toadstool are in no doubt. The active ingredient, as in the Panthercap (plate 33), is myceto-atropine. This reacts with another poison, called muscaradine. These cause various symptoms and can have harmful effects. However, by cooking in water and then draining the water away, the fungus can be eaten without any harmful effects.

Similar species

Apart from the familiar deep red form which seems to feature in many fairy-stories, and which is sometimes thought by country people to bring good luck, there is also a browner variety: *umbrina*. This could easily be mistaken for the Panthercap *A. pantherina*. The difference lies in the lemon-gold colouring just beneath the skin of the cap of *umbrina*, and the warty, concentric volva at the base of the stem. In North America the Fly Agaric appears to be more orange-yellow. Another species similar in colouring is the famous Caesar's Toadstool *A. caesarea*, a southern European species. The light, orange-red cap resembles the yolk of an egg, and emerges from a firm-skinned volva. Gills, stem and rings are a lovely yellow. It grows on sandy soil especially beneath oak. The edible qualities of this species are considered to be superior to that of all other toadstools.

PLATE 32

Amanita pantherina Panthercap or False Blusher
(family Agaricaceae) ☠ **Very poisonous**

Description

The colour of the cap of this very poisonous species is a dark liver-brown to clear, coffee-brown. The skin is sticky, removable and finely grooved around the margin. (A mountain form exists which lacks the grooves. It is to be found in pine woods.) The cap size is usually smaller than that of the Blusher (plate 34) with which it can be confused. The regular rows of pure white warts—the remains of the universal veil—are at first crowded so that the brown colour of the cap is not yet visible. The universal veil splits away evenly from the base of the stem so that, from above, it appears to have been pushed inside the volva. The white, crowded gills become exposed as the partial veil breaks away from the expanding cap, and the smooth ring stands out obliquely from the stem. The flesh remains white and has a smell of turnip or radish.

Habitat

This species is found in mountains as often as in lowlands, both in coniferous and deciduous woods. The soil can be varied. This species does not fruit at any regular time of the year. The time at which it appears may vary with the weather and from year to year. This unreliable occurrence should be borne in mind, since the Panthercap is one of the more poisonous toadstools and is easy to misidentify.

Edibility

This is a poisonous, inedible species. The poison first of all attacks the nervous system. Irritability and organic disorders resemble those caused by the Deadly Nightshade plant *Atropa belladonna*. They can lead to severe illness, and in not a few cases, to death.

Similar species

The Blusher *Amanita rubescens* (plate 34) which resembles it can be recognized by its wine-red colour, by the folds in the ring, and the bulbous, loose-fitting volva. Very similar is the species *A. spissa* (*see* no. 74). This has a grey-brown to silver-grey cap with large, granular warts, a greyish stem and grooved ring, and a bulbous volva covered with grey rings. It commonly occurs in coniferous woods, from June onwards. Although edible, it is best avoided because of the danger of confusing poisonous species with it. A closely similar variety has a pale grey cap, a floury and slightly rubbery skin, and a white stem which grows deep into the needle-carpet.

PLATE 33

Amanita rubescens Blusher
(family Agaricaceae)   **Edible**

Description

This is an edible species of toadstool of the genus *Amanita*, in which are placed some of the most poisonous toadstools. It is not always easy for a beginner to tell one from another. It is therefore best not to include them among those which are collected for the table, since they can occur in many forms and varieties. The cap of the Blusher is a pale, reddish-brown to wine-red, and can even be whitish in colour. The flaky-white to brownish warts cling tightly to the sticky cap, but are easily washed off by rain. They are the pieces of the original universal veil which covered the young toadstool. The very crowded white gills later show reddish markings. The stem is firm, turning a pale wine-red, and carries a white, finely grooved ring. (At first this covers the gills as a delicate partial veil, then tears away from the rim of the cap to form the ring.) The swollen stem-base is encircled by many scales. In section, many dark-stained patches of flesh may be seen where maggots have been feeding. Uninjured specimens will also "blush" by turning reddish where pressed, and this is a good means of identification.

Habitat

The species fruits from June onwards, and is a common occupant of both deciduous and coniferous woods. It is far more numerous in the older woods. Soil conditions may vary.

Edibility

The Blusher is considered to be one of the best edible species. The frequently mentioned poisonous qualities of the skin on the cap are not proved. It is better avoided unless you are an expert, however, as it may easily be confused with the poisonous Panthercap or brown forms of Fly Agaric.

Similar species

The cap of the dangerous *A. pantherina* is nearer to liver-brown in colour with pure white warts and a finely grooved rim. The white, not reddish, stem appears to have been pushed into the thick-lipped volva. It is more slender in build than *A. rubescens*. Brown forms of Fly Agaric can be recognized by the wart-encircled volva, the reddish flesh and a lemon-yellow layer of flesh just under the skin of the cap. The normal form of Fly Agaric has a handsome bright scarlet white-spotted cap (*see* plates 32 and 33).

PLATE 34

Lepiota procera Parasol Mushroom
   (family Agaricaceae) **Edible**

Description
   This is one of the most sought-after agarics. Its handsome shape adorns the borders of woods and glades, and specimens 16 in. tall and 12 in. across are not unusual. The scales on the cap are noticeable and occur in other species of *Lepiota* as well. Gills remain white and are free of the stem. A special feature is the snakelike markings on the stem which appear as the cap opens. The ring which is left as the cap tears open is movable. The stem-base is swollen but has no volva or fringe on it. Flesh is white and flaky, resembling cotton-wool in texture in the cap, but is firmer in the stem, and more fibrous. It does not stain and has a sweet, nutty taste.

Habitat
   This species belongs to light, grassy places in woodlands. Young specimens have a straight, upright stem supporting a bulb-shaped cap which gives them a resemblance to drumsticks standing in the grass. Later, they look like parasols. They can also be seen in parkland and arable fields, sometimes growing on ant-hills. As a rule only a few will be found in one place—it is not usual to find large quantities.

Edibility
   In the button or drumstick stage, this agaric is one of the tastiest of toadstools. Eating it raw (as is sometimes recommended) is not advised. Many people do not enjoy raw toadstools even though they look appetizing. Uncooked, many species leave an unpleasant alkaline taste in the mouth.

Similar species
   The smaller, duller brown *L. rhacodes* looks similar, but on cutting reveals a saffron-brown colour. It grows more sociably and may be found in large numbers in the needle-carpet in pine woods, especially along the borders. It is also edible (plate 36).

PLATE 35

Lepiota rhacodes
(family Agaricaceae) **Edible**

Description
Although very similar to *L. procera* (plate 35), *rhacodes* has a number of distinguishing features. At first the cap is without scales. Soon, a number of dull grey ones appear, to give the cap a shaggy look. The border has an overlapping fringe, and only the centre remains smooth. It does not reach the maximum size of *L. procera*. The white surface of the stem is broken into a brown pattern. The ring is loose and can be moved up and down the stem. The swollen stem-base is so firmly fixed in the soil that, on removal, a large lump of needles comes away with it. A feature which will ensure identity is that a saffron-brick colour appears in the flesh when it is cut or broken. In dry conditions, grown or stunted specimens may fail to show this colour change.

Habitat
This species grows in colonies, especially among needles, on the fringe of conifer woods, or close to the odd pine or fir tree. The dryness of the ground, due to the dense growth of the trees, does not affect it. In fact, damp places —usually good for toadstools—are avoided. This fondness for conifers is not due to any mycorrhizal association with the roots, as is the case with many of the russula and boletus species, but because of the chemical nature of the needle-humus. Like other fungi found in meadows and grass, it feeds on humus.

Edibility
The dry and sometimes woolly white flesh turns saffron-orange when cut, and has a pleasant, nutty flavour. Young specimens are an excellent substitute for *L. procera*, where these are unobtainable, and make a fine dish. This species is highly recommended, especially when picked young, while still in the closed condition, as shown in the cut specimen in the illustration.

Similar species
A similar species is the Giant Parasol (*see* plate 35). There are many other, smaller parasols, some of which have a reddish flesh. One of these, *L. brunneo-incarnata* is poisonous, but is not found in Britain. Parasols can sometimes be mistaken for scaly mushrooms, such as *Agaricus perrarus*. But in *A. perrarus* the scaling is more of a gold-brown and flatter in shape. The ring is fixed and the gills are flesh-red to deep brown. It smells of almonds or aniseed. In Britain it occurs in oak-woods.

PLATE 36

Agaricus silvaticus
(family Agaricaceae)   **Edible**

Description
In spite of its nut-brown cap, this species belongs to the "champignons", as mushrooms are sometimes called. Since the much-gathered Field Mushroom, *A. campestris*, has a white cap, a common error is to assume that all champignons must be white. A more important feature for the collector to notice is the gills. At first whitish, they quickly turn greyish red, finally a violet-brown. The gills are very crowded, free of the stem. Like the ringed stem, the gills come away easily from the cap—but this characteristic is also true of the amanitas and lepiotas (plates 29–36). In *Agaricus*, there is no volva attached to the stem. In the large specimen in the illustration can be seen the dark upper side of the ring where the purple spores have fallen. Many thousands of these may settle on the ring or stem. When broken, the flesh quickly turns a blue-rose or flesh-red. It smells like freshly sawn beech-wood, unlike other mushrooms which smell of almond or aniseed.

Habitat
This mushroom much prefers the needle-carpet of coniferous woods. In young plantations it occurs in numbers along the borders, under the richly branched outer trees, sometimes very thickly, and deeply buried so that only half the rounded cap is visible. It is not found elsewhere, and seems to favour chalky soil. It appears in July.

Edibility
This is an edible species, and since it grows in groups is a rewarding find for the mushroom-hunter.

Similar species
You may mistake this for *Cortinarius glaucopus* (plate 43), but look for the grey-rose to dark violet-brown gills, and see whether a ring is present. A much closer resemblance occurs with *A. haemorrhoidarius*, to be found under conifers. It is larger, and the cut flesh turns a deep blood-red. Twice as large is *A. perrarus* which has golden-brown scales on its cap, a hollow stem and stainless flesh. It smells of aniseed.

PLATE 37

Agaricus silvicola Wood Mushroom
(family Agaricaceae) **Edible**

Description
In the button stage the yellow to milk-white cap has a bell-like outline, often with a peculiar pointed top, so that it resembles a skittle as it sits on its slender, silky white stem. The skin on the cap can break into very fine scales, but it may remain intact and smooth. It has a dull shine and can easily be peeled off. As with all mushrooms (*i.e.*, *Agaricus*) noting the colour of the gills is the best means of identity. At first greyish white they soon take on a reddish tint which then turns to a clear coffee-brown. The hollow stem carries a ring which has angular folds on its underside, round the rim, giving it the appearance of a cog-wheel. Frequently the stem-base is swollen and covered with soil. The flesh of the cap is very thin and blackish in colour. Both cap and stem stain a chrome-yellow when pressed. The smell is a pleasant aniseed, but it is advisable not to rely on this character alone for identification.

Habitat
This species is found both in deciduous and coniferous woods. It prefers soft humus, especially heaped-up needles under a closed tree-canopy, as well as places where brambles and undergrowth occur. These mushrooms appear from May to November and form in groups or rings.

Edibility
This species can be treated as a good eater, but care should be taken that its sweet taste is not lost through over-cooking. It tastes just as appetizing whether in a salad or a cooked dish.

Similar species
The sturdy Horse Mushroom *A. arvensis* (*see* no. 83) which does not grow in woods, looks similar. It is found in fields and commons from June onwards, where the Field Mushroom, *A. campestris* (*see* no. 82), can also be found. The attractive rosy gills on young specimens, and the simple ring, will identify *A. campestris*. When collecting in woods the button stage of the white amanitas can be mistaken for those of *A. silvicola*. As the amanitas are deadly poisonous, great care must be taken when foraging in woods. The *Amanita* genus has a volva on the stem, and gills which remain white (*see* plates 29–30). Also, beware the sticky, poisonous Yellow-staining Mushroom, *A. xanthodermus* (plate 39).

PLATE 38

Agaricus xanthodermus Yellow-staining Mushroom
(family Agaricaceae)
☺ **Mildly poisonous**

Description
This species looks so much like the Horse Mushroom, *A. arvensis* (*see* no. 83) that it has only recently been distinguished from it. The cap, at first white, becomes greyish yellow—especially towards the centre—in older specimens, or in dry weather. The skin of the crown develops fine brown scales. The fine, rose-tinted gills turn greyish red, then violet-brown as the spores ripen. They do not reach the stem. The stem is somewhat swollen at its base, and carries a ring which is smooth on top, flaky underneath, and has a scalloped border. As in other mushrooms, the stem has many root-like hairs attached to its base. When bruised, the cap and stem colour a chrome yellow, as does the flesh, also, when cut. The smell is disagreeable, similar to carbolic, especially when crushed.

Habitat
This abundant mushroom appears to be confined to chalky or limy soils. It prefers the borders of pine woods, growing among bushes or on waste ground, often in tight rings where large clusters may be found. It grows more abundantly in warmer summers, and may last until October.

Edibility
Doubtless the mushroom-hunter, seeing this species, will identify it as a mushroom and may assume that it is edible. However, on cooking, its repulsive carbolic smell will soon reveal its identity. Eating only a few will probably do no harm, but it can cause a considerable stomach upset, and is best avoided.

Similar species
A young specimen could be confused with a young Field Mushroom *A. campestris* (*see* no. 82), due to a similarity in the rose-tinted gills. The latter, however, has a thinner, more simple ring, a pleasant smell, and is confined to fields. The species *A. silvicola* (plate 38), which also bruises yellow, has a sweet scent of aniseed. The gills are greyish, turning reddish brown. It is found in woods. There is a variety *A. xanthodermus lepiotoides* which has a covering of close-fitting scales on the cap which are a dark almost blackish grey.

PLATE 39

Nematoloma (= Hypholoma) capnoides
   (family Strophariaceae)   **Edible**

Description
   This toadstool is easy to find owing to the fact that it grows in clumps.
The pale yellow cap later turns a rusty yellow in the centre, and shows up a
damp grey-ochre border in wet weather. This rim carries the remnants of
the veil which earlier was attached to the stem. This veil, because of its
frailness, does not remain behind as a ring, as in mushrooms (plates 37 and
38). Gills in young specimens are a pale yellow. On ripening, the spores
turn a lilac-grey, then purple-brown, causing the gills to change colour as
well. Those which leave the larger toadstools fall onto the smaller ones beneath and can be picked up with the fingers. In the illustration, the dark
spore-dust from the large specimen can be seen on the caps of the small
ones beneath. The colour of the spores is helpful in identification. A cap
which is cut from the stem and placed, gills downwards, on white paper,
will drop enough spores after a few hours to make a spore print. This will
show the attractive pattern of the gills (*see* p. 27). The upper part of the
thin, curved stems is the same colour as the cap, but deepens to a darker
rust below. The flesh has a mild taste without any trace of the bitterness
which is usually found in the other species of *Nematoloma*.

Habitat
   As early as spring, the stumps of pine trees—seldom any other conifers—
come alive with thick clusters of these fungi. After a break during the summer
months, from September onwards a further crop of toadstools appears
between the loose bark and sap-wood. At first, the fruiting bodies are
covered in fine, web-like veils (cortina). They spread over the stumps and
hasten decay.

Edibility
   This species is recognized as a very tasty fungus. The tough stem should
be discarded.

Similar species
   A very similar-looking species is the green-gilled Sulphur Tuft *N. fasciculare* (plate 41) which has a sulphur-yellow stem and flesh, and greenish
gills in the early stages. It is just as common on stumps of deciduous trees,
especially of felled fruit trees.

PLATE 40

Nematoloma (=Hypholoma) fasciculare Sulphur Tuft
(family Strophariaceae)   **Inedible**

Description

There is very close similarity between this and *N. capnoides* (plate 40). The cap is a clear sulphur-yellow. The gills are at first green. As the spores ripen, they turn the gills a violet-brown and dust the caps of the young toadstools underneath. The fine veil which at first covers the gills finally hangs as a fringe along the rim of the cap. The upper half of the sulphur-yellow stem has a somewhat greenish tinge. (In other species the stem turns an ochre-red.) The flesh is also sulphur coloured and has a very bitter taste.

Habitat

From May onwards clusters of toadstools appear on old tree-stumps, and on uprooted trees in such places as orchards. Even in December, fresh fruit-bodies can be found, often growing in such profusion that they cover the tree-stumps and roots.

Edibility

Because of its intense bitterness, this species is quite inedible. At one time it was considered poisonous.

Similar species

N. capnoides (plate 40) is very similar, but its gills are at first a clear yellow-ochre, then rather more of a smoky grey. It does not have the sulphur-yellow colour of *N. fasciculare* and the flesh has a mild taste without any bitterness. Another species in the same family, *N. sublateritium*, is much larger with a cap two or three times the size, which is a bright brick-red and easily recognized. Young specimens have yellow, hairy scales on their caps. This species, found only on deciduous trees, is also inedible. It could be mistaken for *Pholiota mutabilis* (plate 42) at a casual glance since this toadstool also has a yellow-brown cap and grows in clusters on tree-stumps. However, the gills of *P. mutabilis* are brown, rather than yellow or smoky-grey. The stem is scaly and brown, and clearly ringed. Brownish gills are also found in a deceptive genus *Gymnopilus* which also grows in clusters and has a sulphur coloured cap. Many of its species are inedible.

PLATE 41

Pholiota mutabilis
(family Strophariaceae) **Edible**

Description

The cap of this deciduous woodland species has a damp look with a dark border, in specimens which are in prime condition. As they age, the colour turns a much clearer ochre. The central boss is fox-brown. The smooth skin feels greasy, and slimy after rain. The clear brown gills turn a cocoa-brown as the spores ripen—as happens with *Nematoloma capnoides* (plate 40). The surroundings also turn brown from the fallen spores. There are a number of raised scales on the ringed stem. The flesh of the cap has a brownish tint and a fragrant smell. The stem looks darker and is fibrous.

Habitat

This species has a strong preference for beech stumps and other deciduous trees. From May onwards it turns up as one of the commonest of woodland toadstools. Along river banks, large clusters may be seen growing on alder- or poplar-stumps, making a very attractive sight. An occasional but less vigorous growth of fruit-bodies may be seen on pine wood.

Edibility

P. mutabilis has long been treasured as a table fungus. When cooked, it gives off a strong, pleasant aroma. Although the cap is thin, there are usually so many specimens clustered together that they make a good meal. When mushroom-picking, the tough stem can be left behind. Caps and gills are both used; but it is necessary to remove any soil or adhering wood particles.

Similar species

The numerous species of *Pholiota* occur on the ground or on tree-stumps. Among the many large and often thickly clustered species are those which have a fine yellow colour and a strongly scaled cap. Of these *P. squarrosa* (*see* no. 103) occurs during autumn in thick clusters, growing for preference at the base of living fruit-trees. It is not much use for eating. Other species are found well above ground, on dead branches or trunks.

PLATE 42

Cortinarius glaucopus
(family Cortinariaceae)   **Edible**

Description

This species is chosen for inclusion in this book as one of the examples of the very large genus of *Cortinarius* which consists of more than 100 species. They all commence growth within a web-like veil, the cortina, which spreads across the gills from cap margin to stem. The very distinctive lilac colour of the gills turns with ripening into a fine rust-brown. Falling spores get caught on the remains of the veil, turning this, also, a rust-brown. They can be seen quite clearly, if you turn a specimen over. The cap is greenish at first, then turns a handsome rusty orange, as in the illustration. Colour may also range from a dull, muddy ochre-brown to grey-green. Gills at first a clear lilac soon become brownish. The slightly swollen stem is similar in colour except for the apex which is lilac-blue, and the base which is pale. The inner flesh of the upper stem is lilac.

Habitat

This species appears from August onwards, in wide, closely packed circles, both in deciduous and coniferous woods. There are often hundreds of fruit-bodies pressed against each other, their short bulbous stems buried in the earth. Young and old specimens can be found together. The variation between one specimen and another is surprising, and the odd specimen can sometimes be difficult to identify. This species is amongst the commonest of the genus, and represents one of the sub-divisions (tribe Cortinariae).

Edibility

The edible quality is not very high, although in fact this species is some-times eaten. At best, it can only be used when mixed with other toadstools.

Similar species

There are several species similar in appearance, and their classification requires close attention. One point to note is that none of this genus is poisonous. The similar-looking species of the genus *Inocybe* have many poisonous members, and are recognized by the small, conical cap, and earth coloured or olive-brown gills. Of these, one of the most dangerous is *Inocybe patouillardii* (*see* no. 119) which has a white, then reddish, cap quite different from a *Cortinarius*, but which could be mistaken for a white mushroom. Two further poisonous species are shown in plates 45 and 46.

PLATE 43

Cortinarius mucifluus
(family Cortinariaceae)   **Edible**

Description
   This species represents a sub-genus of the very large group of *Cortinarius*.
In damp weather both cap and stem have a slimy surface. The cortina,
extending from cap margin to stem, looks more like a skin than a veil. The
cap is coloured a slate-grey to yellow-ochre, usually with a violet tinge
around the margin. The centre darkens, and the margin becomes wavy and
ribbed. Broad gills are at first a violet-grey with pale edges, then turn a
brown to rust colour. The slender stem is whitish with a violet tint, and in
young specimens shows a ring (the remains of the cortina). The flesh is
whitish, often violet in the stem, and has only a faint smell.

Habitat
   As a species which is exclusive to deciduous woods, this toadstool is
found on marl or clay soils from summer until autumn. It is one of the
characteristic species of such woodlands.

Edibility
   This is a good, edible species, although fresh specimens covered in their
slippery skin are not pleasant to gather and prepare for cooking.

Similar species
   There is a whole series of similar species, but it is unlikely that these would
ever be confused with poisonous ones. Very similar is the much larger *C.
elatior* (*see* no. 108) which has large grooves on its cap and broad, veined
gills. In the edible *C. trivialis* the stem is encircled with girdle-like swollen
rings. Size and colour are very similar. The species *C. mucosus* is a strong
fox-red, with rust-red gills, and a robust white stem. It is found among
pines on a sandy soil. A further species *C. collinitus* (*see* no. 113) is found in
coniferous woods in the mountains. These toadstools could be mistaken
for another species, *Rozites caperata* (plate 47). Colour and build are very
similar, but *R. caperata* is not a slimy toadstool; it has a flaky cap and a
small, membraneous ring on the stem. Even so, there are one or two species
of *Cortinarius* (in the sub-genus *Phlegmacium*) which have non-slimy caps,
and a more spider's-web-like type of cortina, and which may cause some
confusion in identification. As there are no poisonous species in the whole
genus, any mistake on the part of the toadstool collector would be harmless.

PLATE 44

Inocybe fastigiata
(family Cortinariaceae) ☠ **Poisonous**

Description

When the cap breaks through the soil it looks like a small skittle; then it becomes bell-shaped, about $3\frac{3}{4}$ in. in size, with a central boss on top. Its straw colour may turn to white or brown. Between the radiating cracks in the skin, which reach the margin, the white flesh can be seen. The gills are an olive-green with flaky white edges. The stem, at first white then ochre and finally brown, is slender but firm and long, reaching about 5 in. The elongated lines and scales seen on it later disappear. The cap-margin and stem are at first joined by fine web-like filaments: the cortina. The yellow-white flesh, dull brown in the stem, is fibrous, and has the characteristic spermatic smell of the genus *Inocybe*.

Habitat

This species is recognized by its size, the bell-like shape, and attractive yellow colour. It is nowhere uncommon. From June onwards, it occurs in small groups in deciduous and coniferous woods, in parks, grassy places, and along woodland rides and roads. The fruit-bodies grow exceptionally large in rotting conifer humus or among beech leaves.

Edibility

The recently discovered muscarine in this species has placed it among the very poisonous species. Since even the specialist may have difficulty in distinguishing the different species of *Inocybe*, it would be as well for the toadstool-gatherer to leave them alone.

Similar species

Mistaking this genus for others can easily occur. Generic features are the skittle-shaped cap with a central boss, the silky or slightly scaly skin, ash-grey to olive gills with white edges, and a floury-white colouring in the upper half of the stem. Most species hardly reach $1\frac{1}{2}$ in. across the cap. There are some poisonous species which are very similar to the above, such as *I. praetervisa*, which are distinguished by the shape of the spores. In this species the yellow cap is smaller and less cracked, and the stem is white with a swollen base. The knobbly shaped spores of this species and its close allies are quickly detected under the microscope, and help to distinguish it from the more egg-shaped *I. fastigiata* shown in the plate opposite.

PLATE 45

Inocybe geophylla
(family Cortinariaceae)   ☠ **Poisonous**

Description
The cap is mostly a glistening white (as shown in the extreme left-hand clump in the illustration); sometimes it may be lilac-tinted (right-hand clump; and there may even be yellowish specimens (centre). This can make identification difficult. The cap is skittle-shaped, then more bell-like and has a small central boss. The silky skin is marked with radial lines. Young toadstools have a fine, web-like veil stretching from the cap-margin to the stem. The grey-white gills gradually turn darker to a muddy olive-brown, the borders remaining white. The slender stem is normally a little curved, and is either white or the colour of the cap. The white flesh has a very characteristic sickly smell which is common in this genus.

Habitat
This most widespread of all the inocybes occurs with frequency from summer to autumn, in groups in mixed woodland, mostly in young conifer plantations, but also in grassy places, under the occasional tree and close to bushes. The soil condition is not critical.

Edibility
As in many related species, the poison muscarine is present. Because of its small size, this toadstool is not usually gathered. However, strict warning should be taken not to eat it.

Similar species
Of the many species in this genus which are coloured white, most of them can be identified with the aid of the microscope. The more sturdy looking species are very much like the genus *Tricholoma* (*see* plates 55–59). The most dangerous species is probably *I. patouillardii* (*see* no. 119). The cap and stem are streaked and spotted a brick-red when bruised. It can be confused with *Tricholoma gambosum*. However, this is recognized by its mealy smell, and uniformly white to creamy yellow colour, whereas the gills of *Inocybe* turn olive-grey. Those who wish to gather mushrooms—*i.e.*, *Agaricus*—should bear in mind that they carry a ring. Both *I. patouillardii* and *T. gambosum* grow in similar localities at the same season. They have no ring. Those with a good sense of smell may notice that *Tricholoma* smells of meal and *Inocybe* has a fruity smell, whereas the mushrooms have a smell of aniseed.

PLATE 46

Rozites (= Pholiota) caperata
(family Cortinariaceae) **Edible**

Description
This is the only species of its genus found in western Europe, and was formerly placed in *Pholiota*. Its peculiar feature is the fine-grained layer on top of the cap which resembles a deposit of white hoarfrost. It remains to the last. In the young toadstool, the ochre colour is often mixed with a watery lilac. The small ring on the stem is always visible at first, but later becomes less so. In many specimens the stem carries a number of silky, white girdles, clearly shown in the fallen toadstool in the illustration. Clear, brown and crowded gills join the stem (adnexed) in such a way as to leave a circular groove where they suddenly curve inwards. This is in contrast to the genus *Amanita* (plates 29–34) and the Mushrooms (*Agaricus* plates 37–38) in both of which the gills are free of the stem. The pale brownish flesh of *R. caperata* is firm and juicy, often full of water, and is fibrous in the stem without being tough.

Habitat
This vigorous-looking toadstool is commonly found in pine woods or mixed deciduous and conifer woods, but not on chalky soils. It is a good indicator of sandy areas. Wherever mosses such as *Sphagnum* occur, together with heather and heath-berries, this toadstool is not far away. Growing in groups, it often shares the habitat with *Xerocomus badius* (plate 6), *Paxillus involutus* (plate 12) and the Chantarelle (plate 62).

Edibility
This species is, unfortunately, often maggot-eaten. It has a very pleasant flavour, and can be recommended for use in all sorts of dishes.

Similar species
Some of the non-poisonous species of the genus *Cortinarius* resemble *R. caperata* closely, but they are distinguished by the cortina which is spread between cap and stem, and also by the deep rust-brown gills in the grown *Cortinarius*. An inexperienced collector might be reminded of *Agaricus*. However, these latter do not have the same ochre-yellow colouring, but are rather more white or scaly brown; then, too, the gills are a flesh-rose turning to coffee-brown, and not pale brown as in *R. caperata*.

PLATE 47

Rhodophyllus sinuatus (formerly *Entoloma lividum*)
(family Rhodophyllaceae)   ☠ **Poisonous**

Description

This very distinctive toadstool is recognized by its somewhat twisted growth and its colour. The clear yellow-grey of the—usually—blunt and wavy cap some 8 in. wide, is only occasionally as pale as in the illustration. The skin is slightly sticky, smooth or finely lined, and raised in very tiny scales. The broad and curved gills are at first a peculiar cream or banana-yellow, soon turning a flesh-pink due to the ripening spores. As in *Tricholoma* there is a circular groove around the stem apex at the inner edge of the sinuate gills. This is called a "moat". The stem is strong and firm, but also slender and often twisted, and has a swollen base. Colour is white with greyish tones; the surface is slightly wrinkled and silky. The white flesh of the stem is strongly fibrous, and smells of fresh meal. This character, common in many toadstools, must not be assumed to be definitive but should be taken account of together with the other features described.

Habitat

This species is found exclusively in pure or mixed deciduous woods, more in undergrowth than in open woodland, especially in old beech woods. It occurs in groups or in rows well into autumn, but in variable numbers, on limy or clay soils, but rarely on loose sand.

Edibility

Even eating a small portion could cause severe stomach pains. This species is included among the poisonous toadstools which must be treated with caution, even though death may not result from eating it.

Similar species

This pale and slender species can easily be mistaken for others in the genus, although none of them reach its size. One of these is the mildly poisonous *R. rhodopolius*. It has a silver-grey cap which darkens and becomes watery-looking in wet weather, and does not have the smell of fresh meal. Another, smaller species, *R. nidorosus*, has a sharp smell of ammonia or salt-petre. Both of these are found in deciduous woods and look very similar. Some species in the genus *Clitocybe* resemble *R. sinuatus* (plates 49–51). In the case of *Clitocybe nebularis* (plate 49) the size and colour are very similar. However, the gills are white, then turn yellow not reddish, and are decurrent. *Tricholoma gambosum* has a mealy smell, and decurrent gills. It grows in spring and summer, in rings and rows in grassy places and in meadows, seldom in closed woodland. It is often found in company with various *Rhodophyllus* species all of which have rose-coloured gills and a darker cap.

PLATE 48

Clitocybe nebularis
(family Tricholomataceae) **Edible**

Description

The cap of this common autumn toadstool is a pale grey. The rim is somewhat curved at first, then later spreads out and becomes wavy. The central boss flattens out and the cap becomes funnel-shaped, as is usual with this genus. Small, white, circular patches on the cap are not unusual. They occur wherever something falls on the toadstool, such as dirt or an adhering pine-needle. A useful recognition is the decurrent gills, at first white then creamy yellow, which give the cap a swollen appearance. The whitish stem is grooved in a network pattern. The flesh has a spicy smell, although not very pleasant. At first it is firm but later turns soft.

Habitat

Like the Blewitts and the Blue-leg (plate 61) this toadstool grows in wide and crowded rings. From September onwards the fruit-bodies push upwards out of the deep humus-litter of coniferous or deciduous woods, often growing together with the Blewitts *Rhodopaxillus* (= *Tricholoma*) *personatus*, but in such a way that each species forms its own ring. Humus-dwellers such as these are closely connected with woodlands, and are strongly tied to their habitat.

Edibility

This species is edible and is of a marketable quality. It is, however, best to choose young specimens, as the older ones soon decay. Although this toadstool does not suit every palate it is popular with many people.

Similar species

To distinguish this species from others, the colour and shape of the gills should be examined. The species *C. alexandri* can also be found growing in rings in conifer woods. *C. alexandri* has a thicker and shorter stem, coloured an ochre-grey, a permanent rolled-over edge to the cap, and ochre coloured gills. As it is also edible, no harm can come from a misidentification. This is not the case with *Rhodophyllus sinuatus* (plate 48). This species is poisonous. The cap grows to a width of up to 8 in., and the stem has sinuate, not decurrent, gills as in the genus *Tricholoma*. These are a cream-yellow at first, then flesh-red. The pleasant mealy smell makes this toadstool very tempting —especially as many edible species have the same smell. It is also found in deciduous woods.

PLATE 49

Clitocybe geotropa
(family Tricholomataceae) **Edible**

Description
This toadstool has two special features. Firstly, it can reach a height and breadth of 12 in., and secondly, the stem develops before the cap, as can be seen in the left-hand specimen in the illustration. It has already shot up strongly while the cap is still quite small. In the centre of the cap appears a clearly defined boss. The margin, at first curved over, spreads out, and the cap finally becomes funnel-shaped. The dirty, ochre-coloured gills are decurrent. The stem is the same colour as the cap. The firm, white flesh soon softens and slowly decays. It has a pleasant, spicy smell.

Habitat
This is a striking autumn toadstool which grows mainly in rows or rings in woodlands. The trees and soil vary with the habitat. Best specimens are to be found in places which are poor in other toadstools, in woods which are neglected and overgrown, or damp and weedy spots. A fully grown toadstool can last through until winter.

Edibility
Only the young specimens should be eaten. When fully grown, this species loses its firmness and pleasant taste.

Similar species
Even larger but similar in build is the edible *C. candida*, which is found in grassy places in woods and along their borders. The cap can grow to a width of over 12 in., and is quickly recognized by its striking whiteness. It appears as early as the summer. Growing in rings among the needles of young plantations is *C. alexandri*. It resembles young specimens of *C. geotropa*. Its grey-ochre gills are decurrent on a thick stem which is covered with needles, so that a whole handful of needles comes away as the toadstool is removed. Smaller, but equally edible and just as lasting, is *C. infundibuliformis* (no. 44). It has similar colouring to *C. geotropa*, and appears from June onwards in many forms, both in deciduous and coniferous woods. It can cause a kind of nettle-rash when eaten. Sometimes a young specimen of the poisonous *Rhodophyllus sinuatus* (plate 48) can be mistaken for *C. geotropa*, but may be recognized by its sinuate gills and mealy smell.

PLATE 50

Clitocybe dealbata
(family Tricholomataceae)   ☻ **Poisonous**

Description

In spite of its small size and insignificant appearance, this is a dangerous toadstool. Its 2 in.-wide cap (seldom more) is pure white, slowly changing to a dull yellow. The slightly wavy cap soon acquires a hollow centre, and finally an irregular, wavy border. The skin has a dull lustre. The gills are yellowish, then become a duller ochre, and run down the thin, white stem (decurrent). The very thin flesh has a slightly mealy smell.

Habitat

This toadstool appears in small to average sized rings in meadows and pastures, as well as grassy places. These rings resemble those of puff-balls or mushrooms (*see* p. 20). The soil becomes enriched, producing a ring of bright green grass in which the tiny toadstools may be found.

Edibility

Because of its high muscaridine content this species is placed among the more dangerous of toadstools. It is easily recognizable and is fairly common in Britain.

Similar species

It can be found in company with other white or yellow species. Very similar to the white variety of *C. dealbata* is *C. rivulosa*. It loses its whiteness very rapidly, and the skin becomes cracked in a concentric fashion, a flesh-brown colour appearing in the openings which gives this toadstool an ugly appearance. There is no mealy smell. It grows in similar situations. Away from woods, the edible Fairy-ring Champignon (plate 52) grows in abundance. Its thin cap, although similar in shape, is not white but ochre-yellow; the stem is slimmer, tough and fibrous, and the gills broader and more distant. In grassy places in woods the edible *Clitopilus prunulus* can be seen. It differs from *Clitocybe dealbata* in having more steeply decurrent gills, a flesh-red colour, and a stronger smell of meal. The whole toadstool is more compact and seldom grows in rings, but rather in open groups. Many other similar-looking species of *Clitocybe* are to be found in woodlands; some care should therefore be taken when picking any white specimens for eating. The dull-white *C. dealbata* should serve as a warning not to experiment with an unrecognized specimen, except in the case of those genera in which no poisonous species occur.

PLATE 51

Marasmius oreades Fairy-ring Champignon
   (family Tricholomataceae)   **Edible**

Description
In spite of its smallness, this toadstool is widely known as a delicacy. In warm and damp weather the cap can reach a size of $2\frac{1}{2}$ in. When wet the margin turns a dark and watery colour, and when dry it turns ochre. Some specimens show a pattern of wavy lines. The pale, leathery gills are widely spaced, broad and free from the stem. The stem is so tough that this toadstool cannot be picked unless pulled out of the earth from the very base of the stem. The yellow flesh has a spicy smell, similar to that of cloves. In dry weather it can shrivel up, becoming as hard as horn. With rain it swells up again, and continues to survive.

Habitat
From early spring this toadstool can be found in fields, pastures and grassy places, growing thickly in fairy rings which may measure many yards in diameter. It can also be found in fields where cattle graze, or in soil which has been sheltered by trees for a long period.

Edibility
Although tough and small, this species has a flavouring which is greatly to be recommended. Collecting is certainly tedious, however. The tough stems are unsuitable, as are old specimens which have revived after rain.

Similar species
White species with soft flesh belonging to the genus *Collybia* found in woods could be mistaken for the above. They have crowded gills and a cylindrical stem which breaks easily. Large specimens of *M. oreades* can reach the size of *Marasmius peronatus* the stem of which is distinguished in a striking way by yellow hair-like bristles. The flesh has a strong peppery taste. It has not much value for culinary purposes. *M. oreades* should not be mistaken for *Clitocybe dealbata* (plate 51) the white cap of which soon turns yellow, and the stem of which is much softer and thinner, and the gills slightly decurrent. *C. dealbata* also grows in rings, however, *Marasmius* is a genus rich in species. All of them can recover after rain. Most are very small, and grow on stumps, fallen needles or on leaf-mould. The species *M. perforans* grows in thousands, each slender stem attached to a separate pine needle. The similar-looking *M. scorodonius* is much treasured for flavouring.

PLATE 52

Laccaria amethystina Amethyst Toadstool
   (family Tricholomataceae)   **Edible**

Description

This little toadstool is most striking in its colouring. The gills are thick and wide apart and the flesh of the very thin cap and equally thin, cylindrical stem is somewhat fibrous. The strongly violet coloured gills appear to have been dusted with flour; this is due to the pale lilac spores. When dry, the colour of this toadstool fades except for the gills which remain violet.

Habitat

The Amethyst Toadstool is found from summer into late autumn, especially during wet seasons. It grows in groups in both deciduous and coniferous woods. As a humus species it fruits best in leaf-litter and on moss carpets. It also occurs in clearings, along grassy woodland rides, and in cart-tracks where rainwater gathers. Outside woods it is found in parks and gardens.

Edibility

These toadstools are really not substantial enough to be worth eating, although they have a pleasant taste. In large quantities, they need not be despised for the table.

Similar species

Very similar to the above is *L. laccata* (*see* no. 52), which is a flesh-red colour. A host of small toadstools with thin gills and white spores, of the tribe Collybieae, are related. One of these of similar colouring is *Mycena pura* (*see* no. 58). The gills are pale and crowded, and smell strongly of radish. It is useless for the table. Best for flavouring is the tiny *Marasmius scorodonius* whose dark brown, slender stem is attached to roots or fallen branches. The tooth-bordered cap is reddish brown, the gills white, and it has a strong smell of garlic.

PLATE 53

Armillariella (= Armillaria) mellea Honey Fungus
(family Tricholomataceae) **Edible**

Description

This recognizable species grows in clusters. The cap is a livid yellow, a tawny or dirty brown, and is suffused with green. The surface is downy or scaly, especially towards the centre; the margin is thin, curved and striate. The stem is similar in colour to the cap, but blackens with age. It is spongy inside, and striate above the ring. The ring is white, with a sulphur-coloured border. It is large and persistent. The gills are whitish, and may be spotted brown turning rufous, and are decurrent. The flesh is brownish, and in fresh specimens the taste is acrid and the smell repulsive.

Habitat

This is a harmful parasite of living trees, especially conifers. It often attaches itself to mature pines growing on soil where there is standing water. It is, however, mostly distributed on tree-stumps or on felled trees. It is to be found growing in clusters of about fifty from September onwards. Occasionally in some years, the young fruits will appear as early as July, and may easily be mistaken for some other species.

Because it parasitizes trees, this fungus is sometimes called "the Forester's Curse". The spores are dispersed freely on dead stumps, and the mycelium spreads to produce fresh sporophores. The rhizomorphs, which are made up of many hyphal strands, penetrate underground for long distances. They may contact living tree-roots and force a way in. Reaching below the bark, a rhizomorph flattens out into a web-like mycelium. This spreads in the cambium layer to attack the phloem and xylem wood via the medullary rays, eventually killing the tree. The wood is penetrated and slowly rots away. The mycelial spread shows up as an irregular black line in a section of infected trunk.

Edibility

Young specimens make good eating, but when cooked may become a little slimy. The taste does not suit every palate. Symptoms of poisoning have been noticed in some people. It is advisable to mix these fungi with other species. Where large quantities are found, it is best to preserve these and use them as flavouring.

Similar species

Occasionally species of the genus *Pholiota* are mistaken for the Honey Fungus. These, however, never undergo change in the colour of the gills. Also, the ring, and the scales on the stem, remain brown. They are non-poisonous. One of them, *Pholiota mutabilis*, is an outstanding table fungus (*see* plate 42).

PLATE 54

Tricholoma scalpturatum
(family Tricholomataceae)   **Edible**

Description
The cap changes from a skittle-shape to a more bell-like outline, finally spreading out. As a result, it not infrequently tears at the margin. The width hardly reaches $2\frac{1}{2}$ in. The skin is a silvery to pale lilac-grey, often whitish, the skin breaking into a fine, felt-like surface of fluffy scales. The white gills later show orange-yellow spots. In some cases a yellow tone suffuses the stem, as shown in the illustration. However, this is not a reliable character. The flesh is white and smells somewhat like fresh meal. The taste reminds one of cucumber.

Habitat
This species grows in crowded rows or rings, often many yards across, on grass or needle carpets under pines and firs, often under a single tree in a pasture or woodland glade. It prefers a more neutral humus to an acid one.

Edibility
A close relative, *T. terreum*, can be found in similar situations. This species has a closed, velvety, dark grey cap with fine scales, and grey gills. The flesh has no smell or taste. Both species are edible, and are best mixed with other toadstools.

Similar species
Care should be taken not to confuse *T. scalpturatum* or *T. terreum* with another grey-capped species, *T. pardinum* (plate 57), which is poisonous. It has a powerful growth, a cap-width of up to 5 in., with large, pale grey scales on a pale background. In young specimens, water drops appear at the top of the stem. Like *T. scalpturatum*, it smells of meal. The beginner would be wise to leave these grey toadstools alone. The genus *Tricholoma*, which has about 50 species, contains some good edible kinds. All have in common the fleshy but fibrous stem and the pale, sinuate gills which are covered with a white layer of spore-dust when ripe. Further examples of this genus appear on plates 56–61.

PLATE 55

Tricholoma orirubens
(family Tricholomataceae)   **Edible**

Description
The cap is a remarkable blackish colour with a finely scaled skin. It presents a marbled surface, due to a number of deep grooves. Where it grows in clusters the cap will remain smaller. Whitish gills have a black border, but can also turn completely grey. Old specimens may show attractive rose-coloured patches here and there. These can appear where the specimen has been damaged or where the cap has been eaten into, as in the illustration. However, it is possible to turn up hundreds of specimens without seeing a single sign of redness: it appears to be a feature of the older toadstools. The stem remains white but can show blackish spots or markings. The flesh smells and tastes strongly of flour, and in some cases is rather sweet. Occasionally, there are blue-green patches on the stem which is a specific feature. More typical is the clear yellow at the stem base, only revealed when the toadstool is lifted, and the fibres of its base separated.

Habitat
This species favours chalk soils among pines and firs. In its habitat it can be very numerous, appearing thickly in rings. The author once found in Germany a ring up to 88 yards across, two-thirds of its circumference containing toadstools. In a length of 11 yards there were 2,000 specimens, which meant that there must have been up to 10,000 specimens in this particular ring.

Edibility
T. orirubens is one of the tastiest as well as the most productive species in this genus. One need only choose the correct moment for gathering.

Similar species
Two common species, *T. terreum* (*see* no. 37) and *T. atrosquamosum*, are very similar. They occur in mountains in conifer woods, and differ from *T. orirubens* in having little colouring, a poorly scaled cap, and an aromatic smell of fruit.

PLATE 56

Tricholoma pardinum
(family Tricholomataceae)   ☠ Poisonous

Description
As far as is known, the large genus *Tricholoma* has this one poisonous species. Its scaly grey cap is often similar to that of *T. scalpturatum* (plate 55). However, the scales are larger and its build more vigorous and fleshy. Often the cap is devoid of scales which can make identification difficult. The white gills seldom turn grey or yellow, but exude clear water droplets, as also happens at the top of the stem. The mealy smell can be misleading when collecting, as other agarics, including the edible *Tricholoma* species, also have this scent. Like them, the flesh is white and smells inviting.

Habitat
This species occurs in both deciduous and coniferous woods, especially on chalky soil. It does not grow in rings as frequently as other species, but is mostly to be found in groups so numerous that other species in the vicinity escape the attention.

Edibility
This must be considered a poisonous species. It affects the digestive organs in a similar manner to the species *Rhodophyllus sinuatus* (plate 48). Although not deadly, it is so unpleasant that it is best left alone.

Similar species
This species is often confused with edible kinds in the same genus. It has the same mealy smell as *T. scalpturatum* (plate 55) but is more strongly built and more scaly. Water drops on the gills are also more usual. Another species, *T. terreum* (*see* no. 37), has no strong smell, and the scales on the cap are a dark grey. The gills are inclined to turn grey. It is found in thick rings close to firs, sometimes spreading out from the woodland border, and appearing regularly in pastures grazed by cows or sheep. In similar situations, but far less common, is *T. cingulatum*. It is recognized by a small but distinct ring on the stem. In contrast to the above it prefers sandy ground. Much more easily distinguished from *T. pardinum* is *T. orirubens* (plate 56) which has minute black scales on its cap, and rose-tinted gills.

Note. Since possible confusion may arise, it is advisable not to pick for consumption any of the grey members of this genus.

PLATE 57

Tricholoma flavovirens
(family Tricholomataceae)   **Edible**

Description

The sulphur-yellow cap turns a chrome to greenish yellow, and for a long while the margin remains curved over. The skin is sticky, shiny in wet weather, and peels off. Gills are a striking chrome-yellow. Only at the final stage do they become bleached with white spore-dust. The short stem is buried in the humus and is similar in colour. The white flesh is firm and appetizing, and tastes and smells faintly of meal.

Habitat

This species is commonly found on sandy soils. It fruits rather late in the year. This attractively coloured toadstool nestles in the needle-carpet causing hardly any disturbance to the surface as it grows. It is rarely found far from conifers, and does not appear in deciduous woods. Acid soil is preferred, and specimens may often be found in localized areas where the base-rock is calcareous and the ground covered by a carpet of needles. It is rather rare.

Edibility

This species has earned a reputation as a fine edible toadstool. It is much valued and can be used in any dish.

Similar species

There are other green species which, if hastily examined, could be mistaken for *T. flavovirens*. This could have serious results. It might, at a superficial glance, be confused with the deadly poisonous Deathcap (plate 29). It might also be taken for *T. sulphureum* (plate 59); but this would not be a dangerous confusion. *T. sulphureum* commonly occurs in both coniferous and deciduous woods. It can be identified by the fact that it has a more slender build, a dry, silky and dull-looking cap, as opposed to a sticky one, and by its strong smell of gas. It is inedible. Another species, *T. sejunctum*, has a mealy smell, but there is a bitter taste to the flesh. Gills are pale and the stem white. The cap has dark, radial streaks. It is not worth eating. Features to note in the yellow coloured species of the genus *Phlegmacium*, with which *T. flavovirens* might be confused, are the web-like veil or cortina, which stretches from cap to stem in the early stage of development, the club-shaped stem and the rust-brown gills. None of this genus is poisonous. Remember, too, that there may be misidentification with species of the *Inocybe* group such as the poisonous *Inocybe fastigiata* (plate 45). This, however, has a slender, fibrous stem, olive to earth-brown gills, and a repulsive smell.

PLATE 58

Tricholoma sulphureum
(family Tricholomataceae) **Inedible**

Description

The sulphur-yellow cap with its slightly russet centre, develops a wavy shape and grows to a width of about $3\frac{1}{2}$ in., sometimes splitting. The cap is at first silky, then smooth and is never slimy. It has a dull appearance in dry weather, and does not peel. The sulphur-yellow gills are very thick and sinuate in shape, and leave a groove around the stem. This feature, together with the fleshy stem and white spore-dust, are all characters of the genus *Tricholoma*. The stem also has a sulphur colour, often with a brownish tint; occasionally it narrows at the base. The yellow flesh has a highly unpleasant smell of gas.

Habitat

This species, easily recognized by its sulphur-yellow appearance, is a typical member of the beech wood community, where it appears without fail from early spring onwards in every situation. A strong, more massive form with a pronouncedly russet cap and a sulphur-yellow margin turns up in conifer woods.

Edibility

At one time this toadstool was thought to be poisonous. It is, however, harmless but inedible.

Similar species

It can be confused with *T. flavovirens* (plate 58). Cap and stem look very alike, but *T. flavovirens* has a silky cap, to which needles and humus adhere in dry weather, and white flesh with a strong, mealy smell. The gills are thin and crowded. The same unpleasant smell as in *T. sulphureum* occurs in the inedible *T. inamoenum*. This is not sulphur-coloured but white, with a yellow-ochre centre. Gills are widely spaced. It occurs in conifer woods. Some species of the genus *Nematoloma* may bear a resemblance to *T. sulphureum*, such as the Sulphur Tuft on plate 41. However, this grows in clusters on stumps, and when young has a fine veil covering the green gills. These gills soon change from green to a purple-brown.

PLATE 59

Rhodopaxillus (= Tricholoma) nudus Wood Blewitt
(family Tricholomataceae)   **Edible**

Description
In its early stages the Wood Blewitt is a beautiful, even, violet-blue, with especially attractive gills. Soon, however, the cap changes to a dull red-ochre, the stem and gills also fade, and the colour of the flesh changes from an intense violet to a duller tone. The smell has a peculiar sweetness which is not very pleasant.

Habitat
This species appears in great numbers during autumn. It often occurs in rings, both on a needle-floor and a humus of deciduous leaves. It is a pronounced humus species. The reason for the fairy ring formation, where hundreds of toadstools may appear, is the outward spread of the mycelium in all directions, at the outer border of which the fruit-bodies appear (*see* p. 20). This ring-like spread occurs with many toadstool species, especially on grassland. Grass and meadow plants take on a deep green at the outer edge, to form the familiar ring. The humus-carpet and the atmosphere inside a wood are particularly suited to the Wood Blewitt. Sometimes it occurs in open places, but it avoids heathland. It often turns up in similar localities to *Clitocybe nebularis* (plate 49) which suggests that their requirements are the same.

Edibility
The Wood Blewitt has many culinary uses. However, the sweet taste does not disappear entirely with cooking. It is a clean fungus, seldom attacked by maggots.

Similar species
Species of *Rhodopaxillus* need not be mistaken for the genus *Tricholoma*, if it is remembered that the spore-dust of the former has a dull reddish colouring while that of the latter is pure white. Another easily recognized species is *Rhodopaxillus (= Tricholoma) personatus* (plate 61), the Blue-leg. It also grows in rings, but nearly always in meadows or grassy places during late autumn. Only the stem is coloured violet, at least a greyish tone. It looks more solid and heavy in build than the Wood Blewitt. Productive areas can supply the collector with a welcome harvest at times when other fungi are beginning to die away.

PLATE 60

Rhodopaxillus (= Tricholoma) personatus Blewitt or Blue-leg
(family Tricholomataceae)   **Edible**

Description

The grey-white to brownish cap of this toadstool is usually found tucked away in the late growth of grass and meadow-herbs after their second mowing. The cap can be very regular in shape, with either a level top or an arched centre. Some caps can reach a width of $4\frac{1}{2}$ in. In wet surroundings they feel soft and greasy to the touch. Damp spots give the cap a spotty or marbled look. The margin remains curved under for some while. On turning over a specimen the beautiful lilac of the stem can be seen. The colour is enhanced by the white to watery brown gills. The stem, normally short and fat, but sometimes growing long and thin, swells out at the top. The dull white to pale grey watery-looking flesh is fibrous. The smell is faintly sour and the flesh has a mild taste.

Habitat

Fruiting begins with the first frosts. The toadstools appear in crowded rings in damp meadows, in grassy plots under fruit trees, and in cattle pastures. They can be detected by the rich, dark circles of grass. Low-lying woods close to lakes and streams appear to be a favourite habitat. The fruits may be found from early October to the end of November.

Edibility

The Blue-leg when young is considered a good eater, but should be avoided when it turns brown.

Similar species

This species has now been placed in the genus *Rhodopaxillus* instead of *Tricholoma*, owing to the reddish colour of the spore-dust. The Blue-leg cannot be mistaken for any other toadstool if one remembers the lilac stem, the habitat, and the late fruiting. Even so, it has a close resemblance to the Wood Blewitt (plate 60) when it turns pale, except that the gills of the Wood Blewitt remain lilac. Another edible species is *R. panaeolus* which is rather darker, and has concentric, pale, watery markings round the ribbed cap. Gills are greyish, later turning flesh-pink from the spore-dust. The slender stem is pale grey without the lilac colouring, and the flesh has a faint mealy smell.

PLATE 61

Cantharellus cibarius Chantarelle
(family Cantharellaceae)   **Edible**

Description
This widely known toadstool with its egg-yolk colouring is very noticeable, especially as it occurs in large gatherings. Its most important character
is the forked and branched veins and ridges of the underside of the cap,
which are typical of this family. It grows in many shapes. The cap is usually
thin and loose, becoming more funnel-shaped as its wavy margin grows upwards. In this way the ridges become decurrent. Flesh is firm, white, and has
a strong aromatic smell of spices. Eaten raw it tastes unpleasantly of alkaline
or pepper.

Habitat
The home of this much sought-after toadstool is among pine needles, in
grassy places under conifers, or among beech trees especially in old, mossy
woodlands. It turns up regularly in localities such as these. This has a marked
effect on its value. Fruiting lasts from early June to November.

Edibility
This Chantarelle is suitable for all kinds of dishes. Its merits are its strong
taste, easy recognition, durability and freedom from maggots. In suitable
areas it grows in large numbers, and can be transported over long distances
without bruising or becoming soft. It is one of the most durable species.
The Chantarelle should be cut into small pieces, cooked only a short while.
It should not be eaten in too large a quantity. It is a good idea to store these
mushrooms in a glass jar, so as to arrest drying out, and so that they remain
fresh for future use.

Similar species
A paler, stronger variety may turn up in deciduous woodlands (var.
pallida, plate 63). In conifer areas a more sulphur-yellow sub-species, var.
amethysteus, sometimes appears. It has violet scales and widespread veins.
The fungus which was at one time named False Chantarelle, is really a
separate species, *Hygrophoropsis* (= *Clitocybe*) *aurantiaca*, in a different
genus. It has an orange cap, a brown base to its stem, and real, although
forked, gills, so that a distinction should be easy. It is to be found in conifer
woods where there are fallen trunks and tree-stumps.

PLATE 62

Cantharellus cibarius, var. *pallidus* Pale Chantarelle
(family Cantharellaceae)   **Edible**

Description

Whereas the egg-yellow Chantarelle is well-known and much treasured, this is not the case with this pale form which occurs in beech woods. It is therefore included as a recommendation to the reader. It differs from the normal form in its stronger growth: the cap can reach a width of $4\frac{1}{2}$ in. Instead of being egg-yellow it is a pale yellow to white. The skin tends to break into fine, scaly hairs. Ridges are thick and forked, also a pale yellow. The white flesh is firm and has the same spicy smell as the normal form.

Habitat

The most productive areas for this variety are in half-grown beech woods, particularly on chalk where the normal Chantarelle seldom occurs. It also appears under beech clumps scattered among pine, where it mixes with the normal form. In such situations the pine-wood form may resemble the beech wood kind, except that it will be smaller. In other places the usual strongly built beech wood variety turns up.

Edibility

As appetizing as the ordinary Chantarelle, this pale and larger variety is much more productive. Because of its duller colouring it does not stand out so clearly against the pale brown beech leaves, as does the yellow form against the dark pine needles and green patches of moss. It is usually of good quality as it is seldom attacked by maggots.

Similar species

From above, this pale variety looks much like the Wood Hedgehog (plate 69) with which it is often associated. The spines on the underside of the Wood Hedgehog cannot be confused with the veins of the chantarelle, however. The amethyst form (mentioned in the text opposite plate 62) often resembles this pale beech wood variety, but has a more slender and twisted stem, and ridges with smaller, more widely spaced ribs. Colour of the cap and stem can be a dark sulphur-yellow. In addition, it is confined to conifer woods. Sometimes one may encounter a colourless form (var. *albus*), known as the White Chantarelle. The veins are pure white. It looks so different from the typical form that the collector may hardly recognize it as a Chantarelle.

PLATE 63

Cantharellus tubaeformis
(family Cantharellaceae)   **Edible**

Description

Many collectors who have known and enjoyed the true Chantarelle are surprised when informed that this grey-brown toadstool belongs to the same genus. Turn it over and the same characteristic veining is revealed, even though the colour of the veining is also a grey-brown. A fairly clear yellow appears on the stem which is slender and tubular. The silky, often concentrically marked, cap is wavy and funnel-shaped. Its hollow centre gives the impression that it has been pierced.

Habitat

This is an autumn fungus which may continue to appear until the beginning of winter. It is even to be found frozen at Christmas! Because of its colouring, it is less striking than the handsome egg-yellow species (plate 62). It may appear in large numbers, closely packed in rows or rings, and grows in damp and mossy places in conifer woods. It is indifferent to soil conditions, and may occur on chalk as well as acid ground.

Edibility

This is a lesser-known edible species with a milder taste than the true Chantarelle.

Similar species

A close resemblance is to be found to the species *C. lutescens*. The cap of this fungus is a little browner, the underside more wrinkled, and it has an orange-brown colour with a bluish tint. The stem is clear orange. It has a strong and sweetly aromatic smell, and grows mainly on chalk among conifers, especially in mountain regions. Like *C. tubaeformis*, it is edible; so that even if the two are confused, there will be no harmful consequences. *C. tubaeformis* has only a mild taste and lacks the aroma of the true species. At one time this genus (*Cantharellus*) was included with the gilled fungi, or Agaricales. However, it is now placed separately in its own family, and has close affinities with the fairy-clubs (Clavariaceae).

PLATE 64

Gomphus clavatus
(family Cantharellaceae) **Edible**

Description
The fruit-body first appears as a small, violet, club-shaped object, then soon develops into a top-like shape with a distinct outer rim, later becoming more spoon-like or ear-shaped. The smooth exterior of the funnel has a veined and wrinkled surface, and the inner side a velvety violet to olive-yellow. The veins are powdered yellow by the spore-dust. The white stem lies half-buried in the humus, and has needles and leaves sticking to it. The white, firm flesh, which is somewhat dry, has a slight scent and a mild, at most slightly bitter, taste.

Habitat
From August onwards, even earlier, this toadstool appears, growing in thick rings. It occurs in conifer woods in the mountains, especially in young silver-fir plantations on chalky soil, rarely in deciduous woods.

Edibility
This species, with its white flesh, makes a welcome and appetizing dish. Old specimens become bitter and musty.

Similar species
Similar in appearance and growing in circles is *Clavaria truncata*. It is a yellower colour without the violet, and resembles the small club-shaped stage of *G. clavatus*. The fruit-body is wrinkled but not veined. Although edible it is not a satisfactory substitute. The violet scaled sub-species of *Cantharellus cibarius* (*see* plate 62), with its yellow-veined underside, differs in the fact that it grows in rings. The young *Panus conchatus* bears a similarity; in its early stages it has a violet felt-like cap and stem, genuine but thin and pale-coloured gills, and tough flesh. It always grows in deciduous woods especially on beech-stumps. Later it turns a leather-yellow, and is recognized by its funnel-shaped cap. In some districts there is another species, also shaped rather like an ear. It is called *Guepinia helvelloides*, and can grow to a height of 4 in. It is gelatinous in texture and orange-red. It belongs to the Jelly Fungi, or Tremellales. It grows in groups near tree-stumps and bushes along woodland rides. It is sometimes used in salads, but is not as appetizing as *Gomphus clavatus*.

PLATE 65

Cratarellus cornucopioides Horn-of-Plenty
   (family Cantharellaceae)   **Edible**

Description
It is a surprise during an autumn foray suddenly to come across a cluster of these dark, trumpet shaped fungi among the fallen leaves of a beech-wood. Sometimes thousands of these toadstools appear together. Dark coloured when growing in a damp position, they take on a pale, slate-grey colouring when dry. In contrast to the Chantarelle, the underside is without veins. Instead, it is irregularly wrinkled, giving an uneven surface. The funnel narrows downwards into the stem, which is usually bent and is well-rooted into the ground. This fungus can last for a long while in woods without showing any apparent sign of ageing, except in the outer edge. This becomes black, wrinkled and shrunken with age, then fades when moistened.

Habitat
This species is almost exclusively found in deciduous woods, especially beech. It can also be found in mixed oak and birch areas where there is a shrub layer of hazel. Here and there, it may appear in mossy conifer woods. Soft humus is preferred.

Edibility
The dull appearance of this species does not make it very inviting. However, it does merit some attention. The massive fruiting makes collecting easy, and the taste is quite pleasant.

Similar species
Somewhat rare, but found in similar places, is another species closely resembling the Horn-of-Plenty in colour, but somewhat smaller. A glance underneath will reveal it as a Chantarelle, *Cantharellus cinereus*. It has pale grey, forked veins. Its stem is hollow with a very thin wall. The possibility of confusing the two is not strong, but this kind of resemblance is a good example of how a well-known fungus such as the Horn-of-Plenty can have a "twin" in a different genus. Both these species appear from August on, but are most common in autumn.

PLATE 66

Clavaria aurea
(family Clavariaceae)   **Edible**

Description
This fungus is very different from the normal toadstool with its cap and stem. It is sometimes called the "Coral Fungus" because of its appearance. The very wide-spreading and heavily built fruit-body later changes from golden to yellow-ochre, then finally to a brownish colour. In this condition it is somewhat difficult to distinguish from other species. The white, sour-tasting flesh of adult specimens is marbled with faint lines. Young specimens look rather like cauliflowers in shape with their thick, short white stems, and their as-yet short, crowded branches.

Habitat
This fairy-club occurs in both deciduous and coniferous woods, notably in pine and fir woods on soil which may be mildly acid or very chalky. Its gleaming fruit-body can be seen growing in rows or in circles. The stalks of young specimens are buried deep in the moss. If they grow among brown beech leaves, they can easily be overlooked. Decay quickly sets in, in the older, brown-coloured specimens. These should not be gathered, even though they are bigger and look more tempting.

Edibility
This species is an appetizing fungus, distinguished by its tender flesh, and is to be recommended.

Similar species
Also edible is *C. flava* found in beech woods. It is not as broad, and has elongated branches coloured a lemon to sulphur-yellow. The part of the stem below ground-level is marked with a few wine-red spots. Far less appetizing is the taste of *C. pallida* (plate 68). It is called the "stomach-ache" fungus, with good reason, since it causes severe gastric trouble in most people. It often turns up year after year in the same place. Young specimens are easy to identify with their dull, yellow-ochre to yellow-grey colouring, and the delicate lilac tips to the branches. Another species, *C. formosa* has orange-rose branches with clear yellow tips. Its flavour is not to everyone's taste. One attractive coral-like species not found in this kind of habitat is the small *Calocera viscosa* (*see* no. 182). Its orange-yellow, gelatinous fruit-body always appears among conifers, especially on tree-stumps. It is not edible, and belongs to one of the families of the Jelly Fungi, called the Dacrymyctaceae.

PLATE 67

Clavaria pallida
(family Clavariaceae)   ☠ **Inedible**

Description

From the main stem many branches emerge in a coral-like growth each splitting into two or three endings. With growth these branch-tips lengthen and spread out. Specimens more than twice the height and three times the width of the larger specimens in the illustration are rare. In size and shape, *C. pallida* is much like *C. aurea* (plate 67) except that the stem is less compact, and the branches more open and elongated. This is not an infallible guide to identification, however, since there is so much variation between one specimen and another. Young specimens are not a clear yellow but a paler ochre colour with a tinge of lilac-pink. In older specimens, as with other species, a dirty ochre colour predominates, making identification more difficult. The flesh is white without the faint marbling, and does not change colour. The smell can only be described as unpleasantly sour.

Habitat

This species appears in summer-time, although early autumn is the main season. Entire rings of this fairy-club make a striking picture in conifer woods, young or old, or under deciduous trees. They appear to thrive on chalky soil.

Edibility

Most people, with good reason, look upon this fungus with distaste, for gastric pains and diarrhoea may result shortly after eating it. The effect which this fungus has on people has been traced to poison in the tips of the branches.

Similar species

C. pallida is similar in colour and shape to three other species: *C. flava*, *C. formosa* and *Calocera viscosa* (*see* p. 164).

PLATE 68

Hydnum repandum Wood Hedgehog
(family Hydnaceae) **Edible**

Description

This very common fungus of the woodlands appears unfailingly each year. Due to its colour, the Germans sometimes call it the "Bread-roll Fungus". In time the yellow-ochre colour not unlike that of a bread-roll turns a bright sulphur. The well-formed, wavy cap sometimes becomes joined to its neighbour, to form a surface some 8 in. across. In most cases the paler margin curves upwards, as seen in the larger specimen in the illustration. The skin of the cap is dry and does not turn slimy. It may break into patches but not into scales. This fungus is easy to identify by looking for the brittle, pale yellow spines under the cap, and the pale yellow flesh which slowly turns to ochre-pink. Old specimens become bitter, and must be scalded first if they are to be used in cooking.

Habitat

The Wood Hedgehog is very common, growing on any soils and in all kinds of woodland, and is just as prolific among conifers as among deciduous trees. It is often to be found in beech woods, especially among mature trees where it grows in rows and rings. Due to its bright colour it can frequently be seen from a distance. It grows more slowly than do other fungi, and, due to its durable flesh, may be found well into November.

Edibility

As mentioned above, old specimens should be scalded before cooking. Stand them in scalding water for a few minutes, then drain. The Wood Hedgehog makes a tasty dish if cut into pieces and baked until crisp.

Similar species

A much smaller, thinner and more brittle form occurs in conifer woods (var. *rufescens*). This is an ochre-orange and has a zoned cap. The stem is slim and brittle. It is not quite so common, but it, too, is edible. There appear to be no poisonous species among these spiny toadstools, but a number which are inedible, the flesh being tough and corky. They can occur in brown, rust even sky-blue colours, and have a characteristic smell of chicory, meal or aniseed. They grow in large rings, and are so profuse that, having found one ring, it is quite usual to detect a second, even a third, growing near by even though none were immediately visible.

PLATE 69

Sarcodon (= Hydnum) imbricatus
(family Hydnaceae)   Edible

Description
The underside of this striking toadstool is covered with very short, closely packed spines. They are soft and will break off if touched. This spiny character distinguishes the Hydnaceae from all other kinds, but as a family the Hydnaceae are not well-represented in Europe. The cap of this species can reach a width of 12 in. The upper side, with its firm, spreading and protruding scales, gives the impression of the feathers of a hawk. The pale grey to dull brown stem widens upwards into the cap. Normally the stem is short and sturdy, and frequently grows to one side of the centre. Distorted caps and stems often occur, whereas the colour hardly ever alters. The firm flesh is a grey-brown, fairly dry, and finally turns a little bitter.

Habitat
This species has a marked preference for pine woods, and almost always where it is present a pine tree can be seen in the neighbourhood. The toadstools grow side by side, forming wide rings, especially on sandy soils between heather and heath-berries, or in moss in woodland glades but not too far from trees. It also turns up on shallow, chalky soil where there are pine woods on limestone.

Edibility
It is necessary to plunge these toadstools in scalding water before cutting and preparing them. Stand them for a few minutes, then pour off the liquid. In this manner the bitterness can be removed. Where there are large numbers, it is preferable to pick only the young ones, as they have only a slight bitterness, or none at all.

Similar species
A similar but inedible species is *S. amarescens*. It has a bitter, gall-like taste, is more reddish, and the pointed stem is well buried in the ground. The lower part of the stem is a dark green. It grows under pine on chalk, but is rare. There is a pored species, *Strobilomyces floccopus* which looks very similar from above. The underside has wide, angular pores with grey tubes, and is therefore not a spiny toadstool. This fungus is very palatable, although of little use because of its rarity.

PLATE 70

Trametes gibbosa
(family Polyporaceae)   **Inedible**

Description
The fruits of numerous species of bracket-fungi, of which this is one of the commonest, can be seen on stumps and trunks of deciduous trees. At first it appears as a small swelling which slowly develops into an uneven bracket. This has an irregular shape, varying in size, and very swollen where it is attached to the tree. This fruit-body may measure up to 8 in. across. The velvety surface is white, grey or brown and covered with concentric furrows. During many months of slow growth, algae settle on it to form a green coat, wherever it happens to grow in damp places. The underside is important for identification. It bears radially arranged tubes which are of unequal length and are elongated in cross-section, so that the pores appear oval in shape. The snow-white flesh is tough and corky, and has no particular smell.

Habitat
Beech-stumps are the commonest situations for this bracket which is found everywhere in beech woods. Growth may take place all the year round, but in autumn especially many new specimens are to be found. A mild winter in no way arrests growth. With the coming of warmer weather in spring the brackets become unsightly. They gradually disintegrate as the tree-stumps dry out, or are attacked by insects.

Edibility
This species is inedible but non-poisonous.

Similar species
The species *T. hirsuta* has a smaller and flatter fruit-body, is more greyish in appearance, is rougher, and has dark grey, rounded pores. Another species of bracket fungus, which grows on conifer stumps or on wounds of standing pines, is *Leptoporus albidus*. This has a small but attractive fruit-body of softer texture. The flesh holds a lot of water and is soft and cheesy when dry. It has an unpleasantly sharp and bitter taste.

PLATE 71

Lycoperdon perlatum (=gemmatum) Common Puff-ball
(family Lycoperdaceae)   **Edible**

Description

The various members of the Puff-ball family are not very easy to identify. *L. perlatum* is the commonest species. The rounded top narrows downwards into a short stem which is more or less distinct. At first, the entire puff-ball is a clear white with a covering of hard tubercles which may come off when rubbed, leaving a fine tessellated surface beneath. Cut through, the contents of a young specimen look snow-white. Patches of the cap portion flake away in angular fragments, similar to the way in which the shell comes off a boiled egg when peeled. The stem has more porous yet tougher flesh. As the fungus matures, the outside becomes brownish, while the inside turns greenish. The stem remains white. As soon as the cap-contents dry out, the apex splits open. Next the mass of hyphal threads, or capillitium, inside the fruit-body, dry off, and the olive-green spores break up into the familiar spore-dust. Pressure from outside causes this to puff out of the ragged hole in the tough skin.

Habitat

This puff-ball is a distinctly woodland fungus. It grows in clusters in young pine woods, old beech woods, and mixed woodland where there is undergrowth. Places which are clear of humus are preferred. In deciduous woods the inverted flask-like fruit-bodies grow in spots where the wind has blown away the fallen leaves.

Edibility

This species, like others of its kind, is edible for as long as the contents remain white, and the specimen young. The rather strong smell disappears in cooking.

Similar species

After *L. perlatum*, the next most common species is *L. piriforme*, which resembles it in appearance. It is brownish, somewhat conical-shaped, and grows in clusters on old wood. The mycelium contracts into a solid mass of branches which are attached in a root-like fashion to the stem. In the young button-stage certain toadstools, in particular the Deathcap, when still enclosed in the volva can be mistaken for puff-balls. A cross-section, however, will soon reveal the gills.

PLATE 72

Lycoperdon hiemale
(family Lycoperdaceae)   **Edible when young**

Description

The fruit-body develops in a few days from a tiny round knob into a globular shape with a warty and granulated surface, about $1\frac{1}{2}$–$2\frac{1}{2}$ in. across. It is pronouncedly circular and sits on a narrow base, which is no more than a slight constriction and is not very clearly defined (in contrast to *L. perlatum* plate 72). The rounded body is slightly flattened on top, and covered with floury-looking granules which easily rub off. On a mature specimen, which takes on a yellow tinge, a pinkish tone can often be observed. The pure white flesh is enclosed in a fairly thick, waxy and fragile skin (the peridium). With age the colour changes to a more definite yellow, then olive-brown, becoming more and more soft. When moistened, this olive colouring breaks through (as shown in the two left-hand specimens in the illustration). In dry weather the inside becomes powdery. At the same time the capillitium breaks down. A narrow opening allows the spores to escape when the puff-ball is pressed. Finally, all that remains is the empty bowl-shaped base of the fungus.

Habitat

Like the Field Mushroom this puff-ball occurs in meadows, in fields where cattle graze and in grassy places. As early as the spring, one can see the dark green grass forming the fairy rings. The fruit-bodies appear after a dry spell followed by rain, especially after the second mowing.

Edibility

Young specimens with white flesh are edible. They come up to the standard of the champignon, with which they can be cooked, and may be treated in exactly the same way.

Similar species

Similar in size, colour and habitat is *Bovista nigrescens*. It varies from skittle- to egg-shape and has a smooth, later more tessellated, outer skin which flakes off. The peridium is in two layers. The outer peels away from the inner, much thinner layer, in egg-shell fashion. It finally turns a leaden grey to black, and becomes paper thin. A dried specimen will sometimes free itself from the ground, and be blown about by the wind. Young specimens are edible and tasty. A real giant among puff-balls is the much rarer, but equally white, *Calvatia* (= *Lycoperdon*) *gigantea* which is found in grass, especially in woods. It can reach a diameter of 18 in., and is fully grown in a few days. It can be eaten when young. The same applies to *Calvatia* (= *Lycoperdon*) *caelata* which reaches a height of about 6 in., and is peculiar to mountain meadows. Its white fruit-body is pear-shaped, and has a rough, warty surface.

PLATE 73

Plicariella fulgens
(family Pezizaceae)   **Edible**

Description

This fungus makes a beautiful picture, growing in the deep green moss under the pines and firs which may be dotted with hundreds, if not thousands, of these pale lead to orange coloured cups. At first, they are not very striking, since the closed-up fruit-body shows only the dull, dirty olive exterior. As it opens out into a cup, the bright colour of the slightly wrinkled inner surface is revealed, making the fungus visible from a long way off. The rim is frequently wavy or torn into irregular folds and slits. The outside has a downy surface, and retains the sooty-olive colouring which may fade only a little into a clay colour. As with other cup fungi, the ripe specimen will liberate spores if brought indoors in a closed box and then exposed to the warm atmosphere. The thousands of expelled spores appear in a fine cloud. A single spore is spherical and is a special character of cup fungi; using this as a means of identification is only possible with the aid of a microscope, however.

Habitat

Fruiting commences with the melting of the last snows in March. Later, by mid-May, the fungus is more in evidence. It belongs to a less widely distributed species which appears to be absent from Britain, but in parts of Europe, as in the mountains of south Germany and in Switzerland, it is so abundant that it can be gathered in large quantities. Favourite habitats are moss-covered valleys with half-grown pines, slopes covered with silver fir, and soil where chalk is present. After a season of great abundance where they have appeared in thick gatherings, many years may pass during which hardly a specimen turns up.

Edibility

A tasty fungus which is only worth gathering when abundant. The waxy, brittle flesh means that it must be transported with care. All nature-lovers will surely wish to ensure that this beautifully coloured fungus is protected.

Similar species

There are a large number of reddish coloured cup fungi, although all of the spring-growing species are much smaller than *P. fulgens*. A large and conspicuous species with scarlet cups is *Sarcoscypha coccinea*. It grows on fallen branches, attached by a slender stem. The orange coloured *Aleuria aurantia* (*see* no. 187) appears mostly in autumn along the limestone woodland paths and on bare soil. This species can grow to 4 in. across. Its very brittle cups lie in flat discs. It is very common, edible and tasty.

PLATE 74

Otidea onotica
(family Pezizaceae)   **Edible**

Description
Among the rich family of Pezizaceae are some of the most sombre as well
as some of the liveliest coloured fungi. Most are minute, measuring only a
few millimetres in size; others, which are relative giants, reach some 6 in.
across. This species is one of the latter. Its disc-like fruit-body appears to be
split right the way down one side. The other side is greatly elongated, giving
the species its characteristic ear-like shape, clearly shown in the largest
specimens in the illustration. Its colour, which is orange to egg-yellow, its
considerable size and its bushy growth will identify it without any necessity
for further examination. As with most cup fungi, this species is very fragile.
When collecting it is best to carry it separately. As with *Plicariella fulgens*
(plate 74), it can produce clouds of spores.

Habitat
This handsome cup fungus can be seen both in deciduous and in coniferous woods during summer and autumn. The sensitivity of this species is
such that its growth depends very much on continuous dampness. It is
restricted to woodland and is dependent upon the humus, rather than on
any particular tree (by contrast to fungi such as *Suillus luteus*—plate 8—for
example). However, it has a certain liking for chalky soil, and consequently
it is more common to find it among deciduous trees than under conifers,
since the needle-carpet of the latter acidifies the ground more rapidly.

Edibility
Because of its thin flesh this cup fungus lacks substance, although considered a good edible species.

Similar species
A mistake in identification is very improbable. All similar species are
edible. Of these the smaller, pale brown *O. leporina* (no. 188) found in conifer
woods can be mentioned. It is hardly worth collecting for consumption,
since at the time when it fruits, in the autumn, there are other, more productive fungi available. Among the larger, more circular-shaped cup fungi,
Aleuria aurantia (no. 187) is the commonest. Its flat, disc-shaped cups appear
in autumn on bare soil, and are more darkly orange coloured than *O.
onotica*.

PLATE 75

Gyromitra infula
(family Helvellaceae) **Edible**

Description
The bizarre-shaped cap of this unusual fungus is composed of very irregularly shaped undulating folds which form into two or three raised peaks. This twisted surface is coloured a purple-brown to chestnut-red, turning to cinnamon or ochre-brown when dried. It has a rich lustre. The stem is of irregular shape, pitted and shallow-rooted; it is white with a violet or ochre tint. The skin is fairly granular. The centre of cap and stem is hollow, and divided into chambers by transverse walls. The inner surface is finely granulated.

Habitat
This autumn fungus is rare in many regions. Damp and somewhat open spots in conifer woods, where there is moss, grass or woodland herbs, is a favoured habitat. It occurs in small groups, so that a little cluster can be very conspicuous.

Edibility
Among the autumn-growing species is *Helvella elastica* which resembles the young stage of *G. infula* in shape. However, the folds in the cap are less pronounced, the colour is paler and the stem thinner. Very similar is *G. gigas*. This is a spring fungus of pale cinnamon-brown, very thick-set with a short stem. Since this species also resembles the poisonous *G. esculenta* (no. 190) it is advisable to scald it at first. This last-named fungus occurs in many places where the ground is sandy, and is one of the commonest of spring fungi. Its reddish brown cap, on a short, white stem, has a swollen appearance. It looks quite different from the other species mentioned. Its poisonous properties are described on the next page (plate 77).

PLATE 76

Helvella crispa False Morel
(family Helvellaceae)   **Edible**

Description

The much-folded cap is perched on a ribbed and furrowed stem. The cap is unevenly curved into a saddle-like shape, giving it an extraordinary look. The upper side of these folds is coloured from ochre to white, the underside being a darker nut-brown. The outer framework of the stem is pitted with chambers, and is cartilaginous in texture. Specimens occasionally reach a height of 6 in. When this fungus is suddenly exposed to sunshine, or if the upper side of the cap is lightly rubbed, a fine cloud of minute spores rises from the fruiting surface (*i.e.*, the hymenium). In the sunlight the spores glitter like dust particles.

Habitat

This is an autumn species. It occurs mainly along woodland rides and roadways, growing under bushes where it is kept moist. It also enjoys grassy borders where there are broken stones and where woodland humus has collected. It seldom occurs in the midst of trees, but is rarely seen far away from woodland.

Edibility

In spite of its tough nature this species is considered an agreeably aromatic table food, especially with those who appreciate substance rather than quality.

Similar species

Very similar to the above is *H. lacunosa*. It differs in its greyish black colour and smaller build. When seen in grass its grey appearance could be mistaken at first glance for a worm-cast. The reddish brown *Gyromitra esculenta* (no. 190) which has a swollen cap appears in March and April, growing in sandy places in conifer woods. Dried specimens are non-poisonous. Eaten fresh or undercooked, however, it can be highly poisonous. The cooking-water must be drained off. With *H. crispa* this is not necessary.

PLATE 77

Morchella esculenta Round Morel
(family Morchellaceae)   **Edible**

Description

The elongated or rounded cap is supported on a tall, yellow-white stem. The cap is hollow and the outside honeycombed with irregular cavities. In young specimens these are close-fitting; they increase in size with growth, so that in a fully-grown morel—which can reach 8 in. in height—they may be as much as half an inch wide. The predominant colour is yellow-ochre, with gradations from pale yellow to a dull orange. The ridges are often tinted a light rust, as shown in the illustration. There are, however, some entirely grey-brown and slate coloured forms which do not change with age or growth. The stem is moderately cartilaginous, whereas the ridges on the cap are brittle. This species has a pleasant, spicy smell which, as in other morels, can be detected on the hands after picking.

Habitat

Places where morels occur seem to be very constant, even though the fruit-bodies do not appear every year. The fruiting season is in the middle of April. In two or three days the fruit-bodies are ripe; fresh specimens may continue to appear into May. They occur with regularity under ash and poplar, but are less frequently found under other deciduous trees. The ground must be loose and contain lime or humus. Too much leaf-litter is a disadvantage, but a surface of loose turf is beneficial. Hard soil will produce stunted specimens. Lowland woods, chalky embankments along woodland roads, as well as the grass verges and banks of ditches along highways where fruit-trees grow, are favoured habitats. Ash woods—even the smallest— should be searched regularly. There is, however, no mycorrhizal link between this fungus and the living tree-roots (*see* p. 26).

Edibility and similar species

All morels are delicacies to be searched for. Mistaken identity between species is possible, but causes no harm. Morels are often prepared without first scalding, in spite of the mild dizziness which can result from insufficient cooking. One can hardly call this poisoning, but it would be better to steep the morels in scalding water, removing them after about five minutes. There is a form of *M. esculenta* which is more conical in shape and which closely resembles *M. conica* (plate 79).

PLATE 78

Morchella conica
(family Morchellaceae)   **Edible**

Description

This morel is identified by its elongated and pointed cap, and its brownish colouring. It will normally grow up to 4 in. in height. The elongated raised ribs, and the long cavities in between, are conspicuous, and are more regularly arranged than in other species. The stem shows a clear granulation, as depicted in the illustration. The base of the cap overlaps the top of the stem somewhat in this morel.

Habitat

This species prefers sandy to chalky soil. It often appears in colonies in shaded grass along woodland borders and pathways, as well as on burnt ground. Fruit-bodies can be seen as early as mid-March, and one or two specimens will still be appearing as late as May.

Edibility

This morel, like the species *M. deliciosa* (*see* **Similar species** below) is considered to be among the tastiest and most eagerly sought-after of the morels. Specimens which are obviously old should be avoided as they are hard to digest, and can be unpleasant due to decay having set in. Those with dried-out or mouldy portions should also be left alone. Scalding beforehand, as recommended for *M. esculenta*, is not necessary. This also applies to the tall *M. elata* which is very productive but not so tasty.

Similar species

Similar in size to the above is the spherical *M. deliciosa* which has a more rounded top and less clearly defined ribs. Its colour is olive to pure grey, and rarely also pink-tinged. It grows in similar spots to *M. esculenta* (plate 78) but can also turn up in the middle of conifer plantations. Fruit-bodies are often found with their swollen bases buried in soil which is bare or only slightly covered with needles. Differences between *M. esculenta* and *M. conica* are often so slight that one is tempted to call the first merely a variety of the second. The species *M. elata* first appears in May, and is often taken for a large *conica*. It is distinguished by its three-sided cap, its size and its colouring. The cap is cylindrical in outline, pointed at first. The white on the stem soon turns ochre, and the ribs on the cap finally become black.

PLATE 79

Choiromyces maeandriformis White Truffle
(family Terfeziaceae)   **Edible**

Description

Of all the truffles in Europe this is the commonest—or, at least, the most easily discovered. This whitish, later on ochre-brown to leather-brown, fruit-body is about the size of a fist, has a bumpy surface, and resembles a cleaned potato. Later the skin cracks into separate patches. Inside the truffle, small, white ribbon-like strands penetrate up through the entire flesh, breaking into many separate branches. In between is the leathery grey flesh which ramifies through the interior, and appears circular in cross-section when the truffle is cut. In this is massed the spore-material which is only liberated when the fruit-body matures and splits open. The fresh truffle is firm and solid, and has a peculiar, strong, aromatic scent which becomes most unpleasant as the fungus dies off.

Habitat

The White Truffle grows below ground, with about its top third showing above. It is thus easily overlooked; and it is quite usual for it to be discovered only by kicking against it accidentally with one's shoe. It is found in pine woods; also in young plantations on chalky ground. In an area a hundred yards across, some twenty specimens can be found, half-hidden in the needle-carpet or in firm ground. The truffle with the cut section in the illustration has been dug out of the ground. The one behind it is still buried. Sometimes heavy rain showers will uncover them. Fruiting occurs in summer and early autumn. It helps to know that many truffles live in close contact with certain woodland trees, as is the case with many pored toadstools (Boletaceae) and brackets (Polyporaceae) where a symbiotic link between fungus and tree-roots occurs.

Edibility

The White Truffle is prized as a flavouring for soups, vegetables, meats and sauces. Cut into slices and fried, it provides a very aromatic dish.

Similar species

There is little chance of mistaken identity, if the globular shape, the pale inner colouring, and the marbled appearance of the flesh, is noted. The slightly poisonous earth-ball *Scleroderma aurantium* could be mistaken for a truffle, but this belongs to an entirely different order of Stomach Fungi (Gasterales). It differs in growing above ground, or only slightly buried, and in the violet-black contents of its firm and scaly fruit-body.

PLATE 80

Key to the Genera of Gilled Toadstools

Characters to Examine	White (Leucosporae)	
I Fleshy toadstools—not woody or leathery		
A With a ring or a volva		
1. Both ring and volva	Amanita	
2. Only a volva	Amanitopsis	
3. Only a ring (see 4.)		
4. Gills free	Lepiota	
Gills adnexed or slightly decurrent, cap yellow	Armillariella	
Gills adnexed, cap white	Oudemansiella	
B No ring and no volva		
5. Gills waxy, decurrent	Hygrophorus	
6. Gills exude milk when broken	Lactarius	
7. Stem eccentric or absent	Pleurotus	
8. Stem cartilaginous (see 9.)		
9. Gills decurrent	Omphalia	
Gills adnate or free, cap thick	Collybia	
Gills adnate or free, cap thin, bell-shaped	Mycena	
10. Stem fleshy (see 11.)		
11. Gills free		
Gills sinuate	Tricholoma	
Gills decurrent, thin	Clitocybe	
Gills adnate, distant	Laccaria	
Gills decurrent, easily separated	Paxillus (Lepista)	
Gills adnate, parasitic	Nyctalis	
Gills adnate, flesh brittle	Russula	
C A filamentous or web-like veil in young, which may persist		
12. Veil forming a ring on stem (cortina)		
13. Veil attached to cap		
14. Gills self-digesting from below upwards		
Autodigestion absent		
II Tough or woody toadstools		
15. Stem central. Dried specimen revives when moistened	Marasmius	

Three

(Agaricales) Found in Britain

Colour of the Spores

Pink (Rhodosporae)	Brown (Ochrosporae)	Purple	Black (Melanosporae)
Volvariella			
	Pholiota	Agaricus Stropharia	
			Gomphidius
	Naucoria	Psilocybe Psathyra	Panaeolus Psathyrella
Pluteus Rhodopaxillus } Rhodophyllus }	Hebeloma		Nematoloma
	Paxillus (Tapinia)		
	Cortinarius Inocybe Bolbitius		Coprinus

7

Brief mention has been made, on pages 15–20, of the major groups of fungi, together with their methods of spore production. The four groups may be described as follows:

1. Phycomycetes Of microscopic size. Spores are produced either asexually from a sporangium, or sexually sometimes from a zygo-spore (fig. 1A and B, p. 16). Many exist in water, and produce motile flagellate spores. Some are unicellular, others have branched hyphae without cross-walls. They are usually the province of the micro-scopist, student of pond-life, and the bio-chemist who is concerned with their various useful and harmful effects (*see* p. 21).

2. Ascomycetes The spores, eight in number, are produced in a flask-shaped ascus (fig. 1D, p. 16). Asci are borne on the surface of or contained in the fruit-body which may be saucer-shaped, flask-shaped or tube-like, the last growing underground.

3. Basidiomycetes The spores, mostly four in number, are borne on stalks from a swollen cell, the basidium (fig. 1E). Basidia form a compact layer, the hymenium, which is spread over the fruiting sur-face of the fungus (the gills, pores, spines, etc.). Spores are dis-charged directly into the air, as in the agaric toadstools, or form inside an enclosed skin, as in puff-balls.

4. Fungi Imperfecti A number of microfungi lack a complete life-cycle and cannot be placed with certainty in any one of the three major groups. Spoken of as moulds, some are extremely valuable to man, such as the famous *Penicillium* (p. 21). Reproduction is by means of a conidiophore (Fig. 1C).

The fungi are classified on the binomial system founded by Linnaeus. Each class is further divided into Orders, these into Families, then down to Genera, and finally into Species. When these scientific groups become rather large and unwieldy further sub-divisions may be made. For instance, an Order may be divided into Sub-orders, a Family into Sub-families, etc.

Take as an example the Deathcap *Amanita phalloides*. Here the second word is the descriptive or trivial name for that particular toadstool. Together with some close relatives, such as the Fly Agaric *Amanita muscaria*, and the Blusher *Amanita rubescens*, these make up the genus *Amanita*. Note that these specific (double) names and generic (single) names are written in italics. *Amanita* together with other genera make up the family *Agaricaceae*. This family, together with some others, produce white coloured spores, and are placed by some authorities into the group Leucosporae, meaning "white spored". Together with all the other groups—*i.e.*, pink spored, brown spored, and so on—all these gilled toadstools come under the Agaricales, an order which belongs to the Basidiomycetes.

With the rapid strides in mycology, and the more critical attention given to the minute details of structure, spores, and chemical contents of fungi, there has been some alteration in the naming and grouping of the larger fungi, especially among Continental workers. For this reason the newer classification is given in this translation, but the older and familiar names, long in use in this country, have been retained. For example the well known Beech Tuft in our beech woods is now named in the genus *Oudemansiella*. In the following text, it is written as *Oudemansiella* (= *Armillaria*) *mucida*.

Anyone wishing to confirm the latest classification and naming of the British fungi should consult the check-list which is prepared by the British Mycological Society.

As a supplement to the 80 species with their colour plates on pages 32 to 191 the following is a selected list of species found in Britain, mostly of common occurrence. On pages 192–3 is a key to the genera of the gilled toadstools, or agarics, which form the bulk of the specimens likely to be gathered during a fungus-foray. This key is based on the various visible characters, as well as the spore colouring. The reader may find this of help when making a run-down of the characters in a doubtful specimen.

CLASS BASIDIOMYCETES Fungi with spores developing on a basidium, usually in fours
Sub-class Homobasidiomycetes Fungi with unicellular basidia—most of the larger fungi

ORDER APHYLLOPHORALES Fungi without gills

Families:	1. *Protohydnaceae*	page 197
	2. *Hydnaceae* (Spiny toadstools)	page 198
	3. *Polyporaceae* (Brackets)	page 198
	4. *Fistulinaceae* (Beef-steak)	page 201
	5. *Clavariaceae* (Fairy-clubs)	page 202
	6. *Cantharellaceae* (Chantarelles)	page 203

ORDER AGARICALES Gilled toadstools, or agarics

Families:	7. *Hygrophoraceae*	page 204
	8. *Tricholomataceae*	page 204
	9. *Lentinaceae*	page 209
	10. *Agaricaceae*	page 210
	11. *Rhodophyllaceae*	page 213
	12. *Coprinaceae*	page 213

ORDER AGARICALES (*continued*)

	13. *Strophariaceae*	page 215
	14. *Cortinariaceae*	page 217
	15. *Russulaceae*	page 219
	16. *Gomphidiaceae*	page 222
	17. *Paxillaceae*	page 222
	18. *Boletaceae* (Pored toadstools)	page 222

ORDER GASTERALES   Stomach fungi. The hymenium is enclosed inside the sporophore. The outer wall is called the *peridium*, and the inner tissue which forms the spore-mass is called the *gleba*. In the peridium may also be a mass of hyphae, the *capillitium*

Families:	19. *Sclerodermataceae* (Earth-balls)	page 225
	20. *Lycoperdaceae* (Puff-balls)	page 226
	21. *Geasteraceae* (Earth-stars)	page 227
	22. *Nidulariaceae* (Bird's-nest Fungi)	page 227
	23. *Phallaceae* (Stink-horns)	page 228

Sub-class Heterobasidiomycetes   Fungi with divided basidia, mainly gelatinous species

ORDER TREMELLALES   Jelly Fungi. Mainly growing on wood. Shrunken when dry, but can swell up when moist

Families:	24. *Tremellaceae* (With longitudinally septate basidia)	page 229
	25. *Dacrymycetaceae* (With divided basidia)	page 230
	26. *Auriculariaceae* (With transversely septate basidia)	page 230

CLASS ASCOMYCETES   Fungi with spores developing inside an ascus, usually in eights

Sub-class Discomycetes Cup-fungi. Cup-shaped sporophore, at first closed, then opens, either sessile or on a stalk, the hymenium spread over the upper surface

ORDER PEZIZALES   Pixie-cups or elf-cups. Sporophore opens out into a disc-shape, usually sessile on bare soil, often after fire, occasionally on walls or plaster and mistaken for the dangerous dry-rot fungus (*see Merulius* p. 198)

Families:	27. *Pezizaceae* (Pixie-cups)	page 231
	28. *Helvellaceae* (Morels)	page 232
	29. *Morchellaceae* (True Morels)	page 232

ORDER HELOTIALES   Disc Fungi

| *Families:* | 30. *Geoglossaceae* (Earth-tongues) | page 232 |
| | 31. *Helotiaceae* | page 233 |

Sub-class Tuberales   Globe-fungi. Globe-shaped sporophore develops below ground. Much prized as delicacies but difficult to find (*see* p. 190).

ORDER TUBERALES

Families: 32. *Eutuberaceae*   page 233
         33. *Terfeziaceae*   page 233

ORDER PLECTOASCALES

Family: 34. *Elaphomycetaceae*   page 234

Sub-class Pyrenomycetes Flask-fungi. The asci are enclosed in flask-shaped *perithecia* which are borne directly on the mycelium, or in a fleshy or corky mass, the *stroma*

ORDER SPHAERIALES   Hard, dark-coloured fungi usually on wood or twigs. Asci form in perithecia which open by pores

Family: 35. *Sphaeriaceae*   page 234

ORDER CLAVICIPITALES   Soft fungi, often brightly coloured

Family: 36. *Hypocreaceae*   page 235

Classified List

CLASS BASIDIOMYCETES

ORDER APHYLLOPHORALES
1. Family Protohydnaceae

Fungi with a spreading, flat fruit-body, the hymenium inferior, usually growing on wood.

Genus **Stereum** (Gk. *stereos*, hard) Fruit-body spreading, hymenium inferior, smooth and even, upper surface velvety, flesh woody. Growing on wood. About 15 British species.

1. **Stereum hirsutum** (L. *hirsutus*, hairy) Fruit-body yellow to greyish, in zones with yellow border, hairy, flattened then wavy and curled over, growing in layers; hymenium inferior, bright ochre or pink, smooth; flesh yellowish and leathery. On stumps, logs and palings, January–December. Common. 4 in. diameter.

2. **Stereum purpureum** Fruit-body whitish to greyish, zoned, downy, somewhat overlapping; hymenium inferior, lilac to purple, more ochre when dry, smooth. On dead branches and logs. Said to attack rosaceous trees, causing "silver leaf" disease which appears as a silvery sheen on leaves where the epidermis breaks away causing

air spaces. Branches die back and fruit-bodies later appear on surface, especially on plum. Common. 3½ in. diameter.

Genus **Merulius**   Dry-rot fungus (L. *merulius*, a blackbird)   Fruit-body with hymenium spreading outwards, usually jelly-like to waxy, smooth then becoming folded and pored. On wood. About 20 British species.

3. **Merulius lacrymans** (L. *lacrymans*, weeping)   Fruit-body expanding with hymenium outside, flat on a horizontal surface, bracket-shaped on a vertical one, spongy and fleshy; large pores rusty yellow, spore-dust rust-red, exudes drops of water during growth. Fruit-body is attached to a network of mycelial cords, white at first, turning black, which spread out to attack further timber, even crossing sterile surfaces such as brickwork and concrete. On logs and timber, worked wood in buildings. All year round. Up to 18 in. diameter.

2. Family Hydnaceae

Spiny toadstools. Hymenium composed of spines, teeth, tubercles or wart-like folds, with or without stalk. On logs or soil.

Genus **Hydnum** (G. *hydnon*, a truffle)   Awl-shaped, pointed spines forming a hymenium which is inferior in some species, superior in others. About 50 British species.

4. **Hydnum repandum**   Wood Hedgehog (L. *repandus*, bent backwards) (*see* plate 69).

5. **Hydnum erinaceus** (L. *erinaceus*, a hedgehog)   Entirely white, yellowing with age. Cap spoon-shaped, hanging, warty and without a distinct margin; stem short or rudimentary; spines up to 2½ in. pendulous, straight, equal and crowded. Tree trunks. Occasional. Edible. 13 in. wide, 13 in. tall.
May cause a form of white-rot, especially in oak.

6. **Sarcodon (= Hydnum) imbricatus** (L. *imbricatus*, covered with tiles) (*see* plate 70).

3. Family Polyporaceae

Bracket fungi growing in bracket formation, singly or in tiers, on trees, logs, etc. The hymenium underneath is lined with tubes which have rounded, angular, oval or sinuate pores. Flesh is usually leathery or woody and the sporophore sessile or stalked.

Note: This family was formerly composed in the main of two unwieldy genera; that of **Boletus** (toadstools with pores) and **Polyporus** (brackets on trees, also with pores). Much microscopic investigation in recent years has shown that the former genus, which grows on the ground and is more fleshy, has closer affinities with the agaric toadstools. It is now placed as a family Boletaceae in the Agaricales (*see* page 222). The brackets remain in the family Polyporaceae. Useful visible distinctions for the genera are the shape of the pores. Some of the names used here may be unfamiliar, so the older names have been retained in brackets.

Genus **Polyporus** (Gk. *polus*, many; *poros*, a pore) Fruit-body bracket-shaped, stalked and domed, flesh white, spores pale.

7. **Polyporus sulphureus** (L. *sulphureus*, like sulphur) Cap reddish yellow to orange, paler with age, with undulating, velvety surface, rarely stalked, growing in tiers; tubes sulphur-yellow, pores minute and rounded; flesh pale yellow, then white, may exude a sulphur-yellow liquid when cut, taste bitter. Tree-trunks. Mar.–Nov. Cap 7–8 in.

A wound parasite, may cause heart-rot in tree. Occasionally luminous.

8. **Polyporus caesius** (L. *caesius*, blue-grey) Cap white, then tinged a blue-grey, silky, rarely stalked, growing in tiers; tubes white, pores turning blue-grey when touched, small, unequal and toothed; flesh white, bluish when cut, soft and watery; spores pale blue. Stumps and dead branches, mainly conifer. Mar.–Dec. Common. Cap 4½ in.

9. **Polyporus squamosus** Dryad's Saddle (L. *squama*, scale) Cap ochre, covered with broad, flattened rows of dark scales, fan-shaped, flat and fleshy, growing in tiers on short stems; stem lateral, ochre with dark base; tubes white, turning yellow, decurrent, short; pores minute, then large and angular. Smell strong. Trunks. Common. Edible. Up to 16 in.

A wound parasite causing heart-rot.

Genus **Piptoporus** Fruit-body firm, bracket-shaped, sessile or slightly stalked, skin of cap very thin, lasting for one year.

10. **Piptoporus (= Polyporus) betulinus** Birch Polypore or Razor-strop Fungus (L. *betula*, birch) Cap pale umber, becoming brown, often mottled, rounded or domed behind near the short stalk, skin smooth, thin margin curved over; tubes white and short, pores white,

minute and round; flesh white and soft, then corky. On birch. May–June. Common. Cap 2–12 in.

The name Razor-strop dates from the days when this fungus was used for that purpose. Strips of it were attached to a wooden block, pores uppermost, and surfaced with siliceous earth. The fungus was also used as tinder, for lining insect-cabinets, or as corn-plaster.

Genus **Daedalea** (Gk. *daidalos*, strangely made—*i.e.*, the pores) Similar to Trametes (*q.v.*) but with thicker dividing walls of the irregular, sinuous, or labyrinthine pores. Cap corky.

11. **Daedalea quercina** (L. *quercus*, oak) Cap buff-brown, darker behind, sessile, with smooth, wavy surface marked with concentric furrows, corky; tubes woody and thick-walled; pores pale grey, sinuate and branching, long. On oak stumps. Common. 7 in.

Genus **Ganoderma** (Gk. *ganos*, sheen; *derma*, skin) Cap hard and woody with shiny surface, spores with two layers, the inner one brown, the outer colourless.

12. **Ganoderma** *(= Fomes)* **applanatum** (L. *planatus*, flattened) Cap cinnamon with white edge turning greyish, covered with a hard crust often dusted with rust-coloured spores, flattened and attached by a broad base, in tiers; tubes rust coloured; pores white, browning when bruised. On trunks. Common. 12 in.

13. **Ganoderma** *(= Polyporus)* **lucidum** (L. *lucidus*, shining) Cap pale yellow then blood-red or chestnut, kidney-shaped, surface polished and shiny with grooves, occasionally sessile; stem same colour, lateral, hard and rough; tubes white then cinnamon, pores white, minute and round; flesh white, then reddish, spongy, then corky and woody. Base of trees or on roots. July–April. Common. 7 in.

Genus **Phaeolus** Fruit-body bracket-shaped or circular, without a skin. Flesh brown or yellow, fibrous, spores pale.

14. **Phaeolus** *(= Polyporus)* **schweinitzii** Cap bright tawny-yellow, centre more date-brown, rough and downy surface, growing in tiers; stem lateral, rusty coloured and hairy; tubes greenish yellow, reddening when touched, decurrent; pores broad, angular and irregular, exuding moisture; flesh reddish brown. In tiers, or singly with a central stem, on stumps and roots of conifers. Common. 6 in. Causes heart-rot.

Genus **Trametes** (L. *trama*, the woof—referring to the walls of the pores) Cap woody or corky, sessile; tubes in irregular lengths, radially elongated, with thickened walls, growing on wood.

15. **Trametes gibbosa** (L. *gibbus*, inflated) (*see* plate 71).

16. **Trametes rubescens** (L. *rubescens*, becoming red) Cap unequal shape, white turning red, flattened, velvety and zoned, margin thin; pores white turning crimson when touched, elongated, narrow and sinuate; flesh white turning red when cut, corky. On willow and alder. Frequent. 2 in.

Genus **Polystictus** (Gk. *polus*, many; *stiktos*, punctured) Similar to **Polyporus**, but thinner, with more velvety cuticle, tubes shallower, developing from centre outwards, on wood, in tiers.

17. **Polystictus versicolor** (L. *versicolor*, changing colour) Cap thin and leathery, variously coloured in velvety zones of black, brown, grey and yellow, growing in tiers, sometimes singly in a circular form; tubes white; pores yellowish, small and rounded, then irregular. On logs. Common. 6 in.

18. **Polystictus radiatus** (L. *radius*, ray) Cap a rich brown to dull orange, with yellow margin, rough but velvety surface, in tiers; tubes rust coloured; pores silvery, glistening, minute. On alder. Common. 4 in.

Genus **Ungulina**

19. **Ungulina (= Fomes) ulmaria** (L. *ulmus*, elm) Cap white, turning yellowish, flattened, rough then smooth with a rounded margin; tubes cinnamon with white pores turning yellow, minute and rounded. On old elm. Common. 7 in.

4. Family Fistulinaceae

Cap fleshy, narrowing outwards, sessile. Pores grow separately and do not join together. Spores are pink. On trunks.

Genus **Fistulina** (L. *fistulina*, a small pipe) Cap fleshy and juicy, tubes distinct and separate, spores pink.

20. **Fistulina hepatica** Beef-steak Fungus (Gk. *hepatikos*, liver) Cap chocolate, blood-red, purplish or liver coloured, darkening with age, rounded, thick, fleshy and slimy; stem if present of similar colour, dotted with points; tubes pale turning reddish, separate; pores pale and round; flesh reddish and streaked like beet-root,

fibrous, liberating a red juice; taste bitter especially in young. Tree-trunks, especially oak. Aug.–Nov. Common. Edible when mature, being more tender and less acid. 16 in.

5. Family Clavariaceae

Fairy-clubs, Stag's-horn or Coral Fungi. Fruit-body fleshy or a little leathery, erect, simple or branched and club-shaped. Hymenium even, all over the branches. Mainly terrestrial.

Genus *Clavaria* (L. *clava*, a club) Fruit-body fleshy, simple or branched and club-shaped. Hymenium over entire upper surface of branches. Terrestrial, about 40 British species.

21. *Clavaria (= Ramaria) stricta* (L. *strictus*, close) Fruit-body ochre, tinged wine-red becoming brownish if bruised, branched with ends pale yellow; stem thick and short, tough, with strands of white mycelium at base; branches slender, cylindrical, erect, with curved and pointed ends; smell spicy, taste bitter. Rotten wood stumps. Aug–Jan. Common. 4 in. wide, 3 in. tall.

22. **Clavaria** (=Clavulina) **cristata** (L. *cristatus*, crested) Fruit-body white or mouse-grey tinged pink, branched, fragile; stem short and rounded; branches numerous, irregular, flattened and divided at ends with sharp, pointed tips; flesh white. Woods. July–Jan. Common. Edible. 3 in. wide, 3 in. tall.

23. *Clavaria fusiformis* (L. *fusis*, spindle; *formis*, shape) Fruit-body bright yellow, elongated, spindle-shaped, smooth, hollow, compressed and furrowed, without a stem, densely tufted; flesh whitish; taste bitter. Woods and pastures. July–Dec. Common. 5 in. tall.

24. *Clavaria vermicularis* (L. *vermicularis*, worm-like) Fruit-body white, brittle, elongated and unbranched, somewhat twisted becoming hollow, densely tufted; no stem. In long grass in woods or meadows. Aug.–Nov. Common. 4 in. tall.

25. *Clavaria cinerea* (L. *cinereus*, ashen) Similar to *C. cristata* but more ash-grey with a purplish tinge, branches wrinkled and irregular with toothed ends. Woods. Common. 6 in. wide, 5 in. tall.

Genus *Sparassis* (Gk. *sparasso*, to tear up—from its appearance) Fruit-body fleshy, erect and much branched, branches more or less flattened and confluent. Hymenium smooth and inferior. On ground with mycelium attached to tree-root. 2 British species.

26. **Sparassis crispa** Fruit-body white or ochre, fleshy, erect and much branched, ends flattened out and turned back, tinged yellow at tips, resembling a cauliflower; stem whitish, stout, blackens with age. Attached to tree-root by a cord-like mycelium. Pleasant smell of aniseed. Mainly conifer woods. Frequent. Edible. 12 in. × 12 in.

Genus **Thelephora** (Gk. *thele*, a nipple; *phero*, to bear—after the surface of the hymenium) Fruit-body fibrous or leathery, varied in shape. Hymenium inferior, even, rough or ribbed, the latter covered in nipples (pappilate).

27. **Thelepora terrestris (= laciniata)** (L. *laciniatus*, fringed) Fruit-body rust-brown, turning black, sessile, of unequal shape, scaly, surface covered with rough hairs; hymenium inferior, pale fawn and papillate. Spores brown. Woods and heaths. July–Dec. Common. 2½ in. diameter.

6. Family Cantharellaceae

Fruit-body fleshy or tough, mostly stalked and funnel-shaped, hymenium inferior with wrinkled surface or in folds.

Genus **Cantharellus** (Gk. *kantharos*, a cup) Fruit-body funnel-shaped, hymenium inferior, thick and waxy, gill-like folds, sometimes branched; cap fleshy or membraneous, lobed; stem central; spores white. On the ground or between moss. About 18 British species.

28. **Cantharellus cibarius** Chantarelle (L. *cibarius*, related to food) (*see* plates 62 and 63).

29. **Cantharellus tubaeformis** (L. *tuba*, a pipe; *forma*, shape) (*see* plate 64).

30. **Cantharellus infundibuliformis** (L. *infundibulum*, funnel) Cap yellowish, ashen or sooty, paler when dry, cup-shaped and wrinkled; stem pale yellow with thicker base; gills pale yellow, turning grey, branched. Woods. Common. Edible. 2 in. wide, 2½ in. tall.

Genus **Cratarellus** (Gk. *krater*, a large bowl) Cap fleshy or membraneous, funnel-shaped, stalked; hymenium distinct but adnate to the fruit-body, inferior, smooth, becoming wrinkled; spores white. About 6 British species.

31. **Cratarellus cornucopioides** Horn-of-Plenty (L. *cornu*, horn; *copia*, plenty) (*see* plate 66).

ORDER AGARICALES
7. Family Hygrophoraceae

Gills thick, waxy and juicy, widely spaced, spores white.

Genus **Hygrophorus** (Gk. *hugros*, moist; *phero*, to bear) Cap fleshy, frequently viscid, gills waxy, thick and distant with sharp edges, often branched. Terrestrial. About 70 British species.

32. **Hygrophorus pratensis** Buffcap (L. *pratensis*, in meadows) Cap a tawny-yellow to buff, fleshy, thin along margin, convex then flat, smooth but moist in wet weather, cracked when dry; stem similar but paler colour, broadens towards apex and dilates into cap; gills decurrent, distant, firm and brittle, connected by veins; flesh white. In pastures and downland. Aug.–Dec. Common. Edible. Cap 4 in. stem 2½ in.

33. **Hygrophorus conicus** (L. *conicus*, cone-shaped) (*see* plate 17).

34. **Hygrophorus hypothejus** (Gk. *hypo*, under; *theion*, sulphur) (*see* plate 16).

35. **Hygrophoropsis (=Clitocybe) aurantiaca** False Chantarelle (L. *aurantiacus*, orange-yellow) Cap funnel-shaped, thin and dry; gills real but much forked, fine and narrow, a deep orange; stem thin and elastic; flesh tough and yellow. Conifer woods. Not particularly agreeable.

8. Family Tricholomataceae

Cap and especially stem are fibrous, mostly soft and juicy. Gills thin with sharp edges. Spores white.

Genus **Tricholoma** (Gk. *thrix*, a hair; *loma*, a fringe) Cap convex and fleshy, stem fleshy, gills sinuate. Spores white. Terrestrial.

36. **Tricholoma gambosum** St. George's Mushroom (L. *gambosus*, a swelling—from the shape of the cap) Cap ochre or pale brown, fleshy, convex and undulating, margin curved over with downy fringe. Stem ochre, thick and solid with downy apex marked by the decurrent teeth of the gills; gills whitish, crowded; flesh white, smell and taste of new meal. Pastures and grassy places. April–July. Often appears on St. George's Day, April 23rd. Common. Edible. Cap 3½ in., stem 2½ in. Best gathered when dry.

37. **Tricholoma terreum** (L. *terreus*, earthy—from the colour) Cap mouse-grey or brownish to black, bell-shaped, downy, brittle and

inclined to split; stem white or grey with downy apex; gills white, then ash-grey, unequal length, sinuous; flesh greyish, taste a little bitter. Beech and pine woods. Most of the year round. Common. Edible. Cap 3½ in., stem 2½ in.

38. **Tricholoma rutilans** (L. *rutilans*, turning red) Cap at first covered with purplish red down, soon turning brownish on a yellow background, bell-shaped, dry; stem yellow with purple scales; gills sulphur-yellow, sinuate, broad and crowded; flesh yellow. Pine woods on stumps. Aug.–Nov. Common. Edible but disagreeable. Cap 5½ in., stem 4 in.

39. **Tricholoma sulphureum** (L. *sulphureus*, like sulphur) (*see* plate 59).

Genus **Rhodopaxillus** (Gk. *rhodon*, rose; *paxillus*, a small stake) Fleshy, not slimy, gills sinuate to slightly decurrent, spore dust pinkish.

40. **Rhodopaxillus (= Tricholoma) personatus** Blue-leg or Blewitt (L. *personatus*, masked—referring to the margin of the cap) (*see* plate 61).

41. **Rhodopaxillus (= Tricholoma) nudus** (L. *nudus*, naked) Wood Blewitt (*see* plate 60).

Genus **Clitocybe** (Gk. *klitos*, a slope; *kube*, a head—from the decurrent gills) Cap fleshy, plane, depressed or flask-shaped, with curved-under margin; gills decurrent, sometimes adnate; stem fibrous, elastic, stuffed, becoming hollow. Terrestrial, in woods. About 80 British species.

42. **Clitocybe nebularis** (L. *nebula*, a cloud—from the downy cap) (*see* plate 49).

43. **Clitocybe clavipes** (L. clava, club) Similar to **C. nebularis** but has a distinctly conical cap, grey stem with bulbous base, and deeply decurrent gills. Woods. Suspect. Cap 3 in., stem 2½ in.

44. **Clitocybe infundibuliformis** (L. *infundibuliformis*, funnel-shaped) Cap pale tan, convex but with a central boss and turned-over margin, later more funnel shaped, fleshy, soft and silky; stem pale, thick and firm, base downy white; gills shiny white, decurrent, crowded, soft with pointed ends. In grass. June–Dec. Common. Edible. Cap 3 in., stem 2 in.

45. *Clitocybe odora* (L. *odorus*, fragrant) Cap greenish, fleshy, convex, then depressed, margin curved under, smooth and moist in wet weather; stem similar colour, short and a little bulbous, stuffed; gills dull, decurrent and adnate, a little distant; taste pleasant with strong smell of aniseed. Grows in clusters, Deciduous woods. Aug.– Nov. Common. Edible. Cap 3 in., stem 1½ in.

46. *Clitocybe flaccida* (L. *flaccidus*, soft) Cap tawny to rust, does not fade, smooth, fragile when fresh, flabby when dry, becoming funnel-shaped but without central boss; stem red-brown, elastic, thick and polished; gills whitish turning yellow, deeply decurrent, crowded and narrow, bow-shaped; flesh pale. Woods, often in rings. Sept.–Dec. Common. Edible. Cap 3 in., stem 2½ in.

47. *Clitocybe fragrans* (L. *fragrans*, sweet-smelling) Cap watery, whitish when dry, fleshy, convex then a little depressed, smooth; stem similar colour or yellowish; stuffed then hollow, elastic and smooth; gills whitish, decurrent but adnate, crowded; smell of aniseed. Woods and grass. July–Jan. Common. Edible. Cap 2¼ in., stem 3 in.

48. *Clitocybe geotropa* (Gk. *geo*, earth; *tropos*, turned—from the strongly incurved margin) (*see* plate 50).

49. *Clitocybe dealbata* (L. *dealbatus*, whitewashed) (*see* plate 51).

Genus ***Armillariella*** (L. *armilla*, a bracelet) Genus similar to ***Clitocybe*** but has a ring.

50. *Armillariella (=Armillaria) mellea* Honey Fungus (L. *mel*, honey— from the colour) (*see* plate 54).

Genus ***Oudemansiella*** Cap slimy and thin; gills thick and white; stem thin, somewhat tough.

51. *Oudemansiella (=Armillaria) mucida* Beech Tuft (L. *mucus*, slime) Wholly white and slimy. Cap shiny white, somewhat wavy; gills white; stem slightly cartilaginous. Parasitic on beech. Common. Edible. Cap 3½ in, stem 4 in.

Genus ***Laccaria***. Size smallish; reddish or violet; cap curved; gills thick, distant, slightly decurrent, powdery white.

52. *Laccaria laccata* (Persian *lak*, lacquer—from the colour, resembling gum-lac) Cap carnation-pink or brick-red when wet, more ochre when dry, convex then more flattened, thin and membraneous; stem similar colour, tough often twisted; gills flesh-coloured, turning white, adnate but with a decurrent tooth, broad and distant. Smells of garlic. Woods and heaths. June–Dec. Common. Edible. Cap 2 in., stem 3 in.

53. *Laccaria amethystina*   Amethyst Toadstool (from the colour) (*see* plate 53).

Genus *Collybia* (Gk. *kollubos*, a small coin—from the regular shape of the cap)   Thick, fleshy cap with curved-under margin when young, gills soft, free or adnate, stem cartilaginous, often deep-rooting. On wood or terrestrial. About 80 British species. Non-poisonous.

54. *Collybia radicata* (L. *radicatus*, rooting)   Rooting-shank   Cap greyish to smoky-brown, thin and fleshy, convex then flattened, sticky; stem pale brown, straight and rigid but twisted, stuffed inside, extending into ground in tap-root fashion, usually attached to dead wood or tree-root; gills white, adnexed but with decurrent tooth, somewhat distant; flesh white and elastic. On or near stumps and wood covered with soil. June–Nov. Common. Edible. Cap 5 in., stem 2½ in.

55. *Collybia fusipes* (L. *fusus*, a spindle; *pes*, a foot)   Spindle-shank   Cap reddish brown paling to tan, fleshy, flattened with a slight central boss, smooth and even, dry and often cracked; stem similar colour, fibrous, stuffed then hollow, swollen in the middle, thin both ends, twisted; gills whitish often spotted, adnexed, broad and distant but connected with veins. In clusters on old tree-stumps. May–Dec. Common. Edible. Cap 2 in., stem 4 in.

56. *Collybia maculata*   Cocoa-dust Toadstool (L. *maculatus*, spotted)   Cap creamy or pale buff, spotted and stained a rust-red, fleshy and compact, convex and rather wavy with turned-in margin; stem whitish, spotted, firm and lined, sometimes rooting; gills creamy, often spotted rufous, very crowded; flesh white, thick and firm. Beech and pine woods. May–Nov. Common. Edible. Cap 3½ in., stem 4½ in.

57. *Collybia velutipes* (L. *vellus*, fleece; *pes*, foot—after the velvety stem)   Cap tawny-yellow, a little paler at the margin, convex then flat, fleshy with thin border, irregular, smooth and viscid; stem yellow above, darkening to black at base, tough and twisted, covered with a dense, velvety down; gills white to yellowish, becoming brown, broad and rounded, unequal length; flesh yellowish, watery and soft, taste and smell pleasant. In clusters on stumps, occasion-

ally parasitic. Aug.–April. Edible. Cap $2\frac{1}{2}$ in., stem 3 in. Said to be luminous.

Genus **Mycena** (Gk. *mukos*, a fungus)   Cap membraneous, smooth and thin, bell-shaped, margin never incurved; stem cartilaginous, slender and hollow; gills adnexed or adnate. About 100 British species, mostly small.

58. **Mycena pura** (L. *purus*, clean)   Cap rose-coloured to purple, blue-grey, even whitish, bell-shaped with central boss, striate at margin, brittle and semi-transparent; stem similar colour, tough and hollow; gills whitish, adnate, broad and connected by veins; flesh white, taste and smell of radish. Woods and fields. June–Dec. Common. ☠Suspect. Cap 2 in., stem 4 in.

59. **Mycena polygramma** (Gk. *polus*, many; *gramme*, a line—after the striations on the stem)   Cap fuscous to pale yellow, seldom white, bell-shaped, dry and smooth, with a striate margin; stem silver-grey, rigid but hollow, longitudinally grooved; gills white to pink, broad and adnexed; flesh grey. In clusters on stumps. Aug.–Sept. Common. Edible. Cap 2 in., stem 6 in.

Genus **Marasmius** (Gk. *maraino*, to shrivel—from the way the toadstool dries up)   Cap pliant, dry and thin; stem cartilaginous; gills free and distant. Terrestrial. About 50 British species, some edible, all small, used as flavouring.

60. **Marasmius oreades**   Fairy-ring Champignon (Gk. *oreias*, a nymph) (*see* plate 52).

61. **Marasmius peronatus**   Wood Woolly-foot (L. *peronatus*, covered with hair)   Cap light yellow or pale brick, flabby and wrinkled, flattened with hollow centre and striate margin; stem white with a dense white downy covering at base; gills cream to brown, crowded and free; flesh yellowish. Woods. Common. Cap $3\frac{1}{2}$ in., stem 4 in.

Genus **Omphalia** (Gk. *omphalos*, an umbilicus)   Cap thin, funnel-shaped with curved-over margin; gills decurrent; stem smooth and cartilaginous. About 40 British species, mostly small, on twigs. Of little culinary use.

62. **Omphalia pyxidata** (Gk. *pixidata*, box-shaped)   Cap yellowish red turning pale, marbled with round patches, fleshy and compact, convex then funnel-shaped; stem white then tinged yellow, firm, solid and curved; gills white or pale yellow, decurrent, distant;

flesh white, smell and taste pleasant. Grass, lawns and woods. July–Nov. Common. Cap and stem 1½ in.

63. **Omphalia umbellifera** (L. *umbella*, shade; *fero*, bear)   Cap deep brown to yellow, convex, then flattened, finally a deeper umbrella-shape with striate margin, silky when dry; stem similar colour, tough; gills white or yellowish, distant, broad and triangular. On damp ground in shady places. Common. Cap and stem 1 in.

Genus **Nyctalis** (Gk. *nux*, night—from growing in dark places) Cap fleshy, gills adnate, distant and thick. Parasitic on other agarics. 2 British species.

64. **Nyctalis parasitica** (Gk. *para*, beside; *sitos*, food)   Cap pale brown to whitish, bell-shaped with slight central boss, with a grey and persistent skin; stem white with a hairy base; gills white then brownish, adnate later twisted and joining. Parasitic on **Russula adusta, Lactareus vellereus** and others.

65. **Nyctalis asterophora** (Gk. *aster*, star; *phora*, producing)   Cap white, then fawn, conical and somewhat woolly; stem white then brown and hairy; gills white, shallow and forked, adnate. Parasitic on **Russula nigricans** and others. Cap and stem ¾ in.
This species produces star-shaped chlamydospores on the cap. These can give rise to the above basidial form.

9. Family Lentinaceae

Genus **Pleurotus** (Gk. *pleuron*, a side; *ous*, an ear)   Cap fleshy, eccentric. Stem lateral or absent. In some species there is a ring. Gills decurrent, sinuate or adnate. Usually on trees. About 40 British species.

66. **Pleurotus ostreatus**   Oyster Mushroom (L. *ostreatus*, oyster-shape)   Cap blackish when young, becoming pale brown or slate-coloured, yellow when old, soft and fleshy, shell-shaped, smooth and moist; stem white, short or absent, ascending obliquely on one side and spreading into cap; gills white may turn yellow, broad, decurrent, joining together near stem; flesh white, taste and smell pleasant. Spore-dust lilac. On stumps, logs and tree-trunks. Jan.–Dec. Common. Edible. 5 in. across.

67. **Pleurotus ulmarius** (L. *ulmarius*, belonging to the elm)   Cap yellowish red turning white, marbled in round patches, fleshy and compact, regular, convex and disc-shaped; stem white, then tinged

yellow, thick, firm and solid, curved and to one side; gills whitish or pale yellow, broad and crowded, adnexed, rounded behind; flesh white, smell and taste pleasant. On trunks, especially elm. June–Dec. Common. Edible. 5 in. across.

10. Family Agaricaceae

Genus *Amanita* (an old Greek name based on the mountain *Amanos* where mushrooms were probably gathered) The universal veil leaves behind a volva at the base of stem, spots on the cap, and most species also have a ring. Gills free; stem bulbous, at first stuffed, then hollow; spores and gills white. Contains some very dangerous species.

68. *Amanita phalloides* Deathcap (named after the Stinkhorn, *Phallus*, which has a similar volva) (*see* plate 29). ☠

69. *Amanita verna* Fool's Mushroom (L. *vernus*, occurring in spring) Entirely white. Cap globular at first, then expanded, thin and viscid, usually naked, with margin finely striated due to gills showing through; stem fairly long and thin but progressively bulbous downwards; ring superior, membraneous, striated; volva thick with a free-lobed margin. Rather uncommon in woods. Name is misleading as it fruits in autumn. ☠**Deadly poisonous**. Cap 3 in., stem 5 in.

70. *Amanita virosa* Destroying Angel (*see* plate 30). ☠

71. *Amanita citrina* (=*mappa*) False Deathcap (L. *citrinus*, lemon-coloured) (*see* plate 31). ☺

72. *Amanita muscaria* Fly Agaric (L. *musca*, a fly—from the old method of killing flies) (*see* plate 32). ☠

73. *Amanita pantherina* Panthercap or False Blusher (L. *pantherinus*, spotted like a leopard) (*see* plate 33). ☠

74. *Amanita spissa* (=*excelsa*) Cap a smoky brown to grey, covered with ash-coloured warts; stem white, rather short and striate above the ring, bulbous and concentrically scaly below. A smell of radish. Woods. ☺Suspect. Cap 5 in., stem 5½ in.

75. *Amanita rubescens* Blusher (L. *rubescens*, turning red—from the change of flesh colour when bruised) (*see* plate 34).

Genus *Amanitopsis*   This genus is now joined to *Amanita* by some authorities, and differs only in the absence of a ring. The cap margin is striated.

76. *Amanitopsis vaginata* (L. *vagina*, sheath)   Grisette (a name also given to the women who sold toadstools in the old Paris markets). Cap a mouse-grey with deeply striated border, and white or grey patches of the broken universal veil, sometimes a whole piece on top resembling a hood; stem long, no ring; volva white or grey, elongated and sheathing, well buried in soil. Woods. Aug.–Oct. Common. Edible. Cap 4½ in., stem 6½ in.

77. *Amanitopsis fulva* (L. *fulva*, brown)   Tawny Grisette. Similar to no. 76 but more brownish. Woods. Common. Edible. Cap 3½ in., stem 6 in.

Genus *Lepiota* (Gk. *lepis*, a scale)   Cap usually scaly with central boss, gills white, free. Stem hollow with a movable ring, no volva. Mainly in grassy places and wood borders. About 50 British species.

78. *Lepiota procera*   Parasol Mushroom (L. *procera*, tall) (*see* plate 35).

79. *Lepiota rhacodes* (*see* plate 36).

80. *Lepiota granulosa* (L. *granulosus*, granular)   Cap rust-brown, becoming pale when dry, at first convex then flattened with a low, central boss, scurfy or granulated, often wrinkled, the margin showing remains of the veil; stem white at top, more brownish with tiny scales below; ring membraneous and torn; flesh yellowish but reddens, especially in lower stem, taste pleasant. Conifer woods. July–Nov. Common. Edible. Cap 2 in., stem 2½ in.

Genus *Pluteus* (L. *pluteus*, a movable house—from the resemblance of the cap to a turret)   Cap fleshy and regular; stem fleshy; gills free turning pink; no ring or volva. Mostly close to tree-trunks. About 15 British species.

81. *Pluteus cervinus* (L. *cervus*, a deer—from the colour of the cap)   Cap fawn, fleshy and fragile, expanded with a central boss, skin viscid in wet weather, finally breaks into patches; stem white, patterned with black fibrils, solid and bulbous, may be curved; gills white then flesh colour, free, rounded and crowded; spores pink; flesh white and soft. On stumps, fallen trees and in saw-dust. Jan.–Dec. Common. Edible. Cap 6 in., stem 5 in.

Genus *Agaricus*, formerly *Psalliota*, the typical mushrooms. Similar to *Lepiota*. Cap fleshy, stem fibrous with a ring but no volva. Gills free, whitish at first, turning pink then deep purple-brown. Spores similar. About 20 British species, all terrestrial, mostly edible.

82. *Agaricus campestris* Field Mushroom (L. *campestris*, belonging to a plain) Cap white to brownish, fleshy, rounded then flattened, dry and silky, sometimes a little scaly; stem white, stuffed, firm and short; ring white, membraneous, median and torn; gills whitish, then pink to purplish brown, free and crowded; flesh white becoming more reddish, soft, smell and taste pleasant. Pastures. May–Dec. Common. Edible. Cap 5 in., stem 3¾ in.

83. *Agaricus arvensis* Horse Mushroom (L. *arvensis*, in cultivated fields) Cap whitish, then stains yellow, fleshy, bell-shaped then flattened, dry and slightly silky; stem white, often stained yellow, hollow with a woolly centre, stout and thick at base; ring white, superior appearing double, upper part membraneous, lower part thick but smaller and radially split; gills white then reddish to brown; flesh white or tinged yellow, firm and juicy, then softens. Meadows, often in rings. Common. Edible. Cap 8 in., stem 4 in.

Both this and the Field Mushroom are very tasty when young. There are many growth-forms.

84. *Agaricus bisporus* (L. *bis*, twice; Gk. *sporos*, seed) Cap fawn to brownish, slightly scaly, with incurved margin; stem stout; ring membraneous; gills crowded, pale pink with two-spored basidia; flesh faintly pink. Roadsides and manure beds, but not usually in grass. Cap 4½ in., stem 2½ in.

The commonly cultivated mushroom probably derives from this species.

85. *Agaricus bisporus*, var. *albida* Cultivated Mushroom (L. *albus*, white) Cap smooth, whitish, pale brown in centre, gills a dirty pink to deep brown. Less tasty than the wild Field Mushroom. (For a description of mushroom-growing, see page 23.)

86. *Agaricus silvicola* Wood Mushroom (L. *sylva*, a wood) (*see* plate 38).

87. *Agaricus silvaticus* (L. *sylva*, a wood) (*see* plate 37).

88. *Agaricus xanthodermus* Yellow-staining Mushroom (Gk. *xanthos*, yellow; L. *derma*, skin) (*see* plate 39). 🍄

Genus **Volvariella** (L. *volva*, a wrapper) Similar to *Amanitopsis*, but with pink spores. Cap fleshy and regular; stem with a membraneous, loose, sheathing volva; gills free. Terrestrial. About 10 British species.

89. **Volvariella (= Volvaria) volvacea** (L. *volvaceus*, having a volva) Cap ash-coloured, streaked with fibrils, bell-shaped, curved-under margin; stem white and solid; volva whitish, elongated, membraneous; gills white then flesh-coloured, free; flesh white. Roadsides and greenhouses. Can be a nuisance in hot-houses. Not common. July–Oct. ☻Suspect. Cap 5 in., stem 7 in.

11. Family Rhodophyllaceae

Genus **Rhodophyllus** (= **Entoloma**) (Gk. *entos*, inside; *loma*, a fringe) Similar to genus **Tricholoma** but spores are pink. Cap fleshy with margin incurved at first; stem fibrous or fleshy; gills sinuate or adnate, turning pink. Terrestrial, a number poisonous. About 30 British species.

90. **Rhodophyllus sinuatus (= Entoloma lividum)** (L. *lividus*, leaden; *sinus*, a curve) (*see* plate 48). ☻

12. Family Coprinaceae

Genus **Coprinus** (Gk. *copros*, dung—a common habitat) Cap membraneous or fleshy, scaly, oval, with margin at first pressed to stem; a ring may occur; gills free or adnate, thin. Autodigestion occurs. On dung, made-up ground, decaying wood, in clusters or solitary. About 50 British species.

91. **Coprinus comatus** Shaggy Inkcap or Lawyer's Wig (L. *comatus*, hairy) (*see* plate 28).

92. **Coprinus atramentarius** Common Inkcap (L. *atramentarius*, inky) Usually bigger than **C. comatus**. Cap grey with a silky texture, fleshy and scaly towards centre, oval and deeply grooved, slightly uneven margin; stem white, easily removed; gills white then deep brown, free; flesh sooty, taste mild. In clusters, occasionally solitary, in woods, pastures, gardens and made-up ground. May–Dec. Common. Edible. Cap 3½ in., stem 7 in.

This species has been used to provide the ink made in the past for pen and brush-work. One method was to boil it with a little water, adding cloves to prevent any mouldiness. Gum-arabic was also added to the ink. Ink made in this way tends to brown but can last for many years, although it easily washes off. It was once used to prevent counterfeiting, since its genuineness could be verified by identifying the spores under a microscope.

93. *Coprinus micaceus* (L. *micaceus*, glittering)   Cap yellowish or leaden, darker in centre, finally date-brown, oval then bell-shaped, lobed and split around margin, covered at first in glistening particles, then smooth and grooved; stem white, hollow, soft and silky, curved; gills white then brown along edge, finally black, adnexed; flesh pale. Dense clusters on stumps and old wood. Jan.–Dec. Common. Edible. Cap 3 in., stem 7 in.

94. *Coprinus plicatilis* (L. *plicatilis*, folded)   Cap brownish, then blue-grey, bell-shaped then expanded into furrows, smooth; stem pale and smooth; gills creamy then grey to blackish, free of stem but joined to a collar formed by the swollen stem apex; flesh white. Woods, pastures. April–Dec. Common.

Genus *Psathyrella* (a diminutive of *Psathyra*—*q.v.*)   Tiny toadstools similar in build to *Mycena*. Cap membraneous, striate with margin pressed to stem; gills adnate or free; spores black. On rich ground or old wood. About 12 British species.

95. *Psathyrella disseminata*   Cap white or yellow turning ash-grey, smooth and grooved; stem fragile, curved, rough then smooth; gills white then black, adnate. Crowded in clusters on old stumps. Common. Cap ½ in., stem 2½ in.

96. *Psathyrella gracilis* (L. *gracilis*, slender)   Cap grey, tan or rosy, watery in appearance, pale when dry, bell-shaped with striated margin; stem white, firm and straight, smooth; gills white then blackish, edge rosy, adnate. Woods. Common. Cap 2 in., stem 4 in.

Genus *Psathyra* (Gk. *psathuros*, friable)   Similar in build to *Mycena*, but with purple spores. Cap fleshy or membraneous, margin at first pressed to stem. Stem fragile, gills pale becoming purple, adnate or free. Terrestrial or on rotting wood. About 20 British species.

97. *Psathyra corrugia* (L. *rugare*, to wrinkle)   Cap rose- or flesh-coloured, paler and wrinkled when dry, minutely scaly; stem white or brown; gills white becoming violet then black with white edge, adnate. Woods. Common. Cap 2 in., stem 3 in.

98. *Psathyra fatua*   Cap tan or clay-coloured, pale and wrinkled when dry, covered in fibrils when young, smooth later, margin tending to break up; stem shiny white, smooth, downy near base; gills white then brownish, adnate and crowded, edged in white. In clusters on rotten wood, in gardens, etc. Common. Cap 1 in., stem 3½ in.

Genus *Panaeolus* (Gk. *panaiolos*, variegated)   Agrees with *Collybia* but species are small and spores black. Cap fleshy, conical not expanded; gills adnate, unequal length. In rich grass or on dung. About 11 British species.

99. *Panaeolus campanulatus* (L. *campana*, bell)   Cap brownish or soot-coloured, slightly sticky and shiny when wet, conical without expanding, the margin exceeding the gills; stem white, then brown, straight and stiff; gills adnate and crowded, varying from grey to black due to dark spores, with white edge showing drops of water; flesh reddish. On horse-dung and in rich grass. Common. Cap 1½ in., stem 4 in.

13. Family Strophariaceae

Genus *Stropharia* (Gk. *strophos*, a sword belt—referring to the ring) Similar to *Armillariella* but with purple-brown spores. Cap scaly and viscid, fleshy; ring membraneous; gills adnate. Mostly in woods, some poisonous. About 20 British species.

100. *Stropharia aeruginosa*   Verdigris Toadstool (L. *aeruginosus*, covered in copper-rust)   Cap with a bluish slimy covering, often flecked with white which may come off to reveal a brownish yellow skin, at first bell-shaped then more flattened with a central boss and curved-under margin; stem brownish covered in white, a little scaly below ring, smooth above; ring white above and woolly; gills white then brownish to purple, edged in white, adnate; flesh bluish, smell strong. Woods and pastures. Common. ☞Suspect. Cap 1–3 in., stem 2–3 in.

Genus *Nematoloma* (= *Hypholoma*) (L. *hyphe*, a web; *loma*, a fringe)   Similar to *Tricholoma* but with purple-brown spores. Cap fleshy with curved-under margin, gills at first covered by a cobweb-like veil. Stem fleshy or fibrous usually yellow to sulphur-brown. Ring represented by fine, dark coloured hairs around the stem and on cap margin; gills sinuate or adnate. On wood, usually in dense clusters. About 25 British species.

101. **Nematoloma (=Hypholoma) fasciculare** Sulphur Tuft (L. *fascicularis*, tufted) (*see* plate 41).

102. **Nematoloma (=Hypholoma) capnoides** (*see* plate 40).

Genus *Pholiota* (Gk. *pholios*, a scale) Scaly cap similar to **Armillariella** but with brown spores. Cap fleshy; stem with spreading membraneous ring; gills adnate becoming decurrent. On wood or terrestrial. About 40 British species.

103. **Pholiota squarrosa** (L. *squarrosus*, rough) Cap saffron to ochre, rough with crowded small scales, bell-shaped with curved-under margin; stem yellowish covered with small, darker scales up to the ring, smooth above and stuffed; ring colour of scales; gills a rust-brown, adnate with decurrent tooth, narrow and crowded; flesh pale yellow, smell heavy and unpleasant. In clusters on old stumps. Common. Sept.–Nov. Edible. Cap $4\frac{1}{2}$ in., stem 4 in.

104. **Pholiota spectabilis** (L. *spectabilis*, notable) Cap golden yellow, fleshy and compact, rounded with central boss, dry and shiny, covered with scales which extend as a veil along the incurved margin; stem sulphur-yellow, firm and extending into a root which is sheathed in a veil, somewhat scaly up to the ring; ring yellowish, rather low; gills yellow then rust-brown, adnate, narrow and very crowded; flesh sulphur-yellow, reddens if touched, smell pleasant, taste bitter. In clusters on stumps. Aug.–Dec. Common. Cap 6 in., stem 6 in.

Genus **Bolbitius** (Gk. *bolbiton*, cow-manure—from the usual habitat).

105. **Bolbitius fragilis** (L. *fragilis*, brittle) Cap pale yellow, closely conical then spread out, smooth, sticky with striate margin; stem yellow, hollow and fragile, smooth; gills yellow then cinnamon, adnexed or just free; flesh yellowish. In clusters on horse manure and other dung. April–Nov. Common. Cap $2\frac{1}{4}$ in., stem 4 in.

Genus **Psilocybe** (Gk. *psilos*, naked; *kube*, a head—from the smooth cap) Similar to **Collybia** but has purple spores. Cap fleshy and smooth, margin at first uncurved; stem cartilaginous, often rooting; gills dark, adnate or adnexed. Small size, usually terrestrial, sometimes in clusters. About 30 British species.

106. **Psilocybe semilanceata** (L. *semi*, half; *lanceolatus*, spear-shaped) Cap yellow-green or brownish, conical and does not expand, slimy; stem pale and smooth with an inner pith; gills creamy

then purple, adnexed and crowded; flesh white. Pastures, waysides. Aug.–Dec. Common. ⊛Suspect. Cap $\frac{1}{2}$ in., stem $\frac{1}{4}$ in.

14. Family Cortinariaceae

Genus **Cortinarius** (L. *cortina*, a veil)   Edge of cap joined to stem in young stage by a web-like veil, the cortina. The universal veil may also leave a ring on the stem. Cap fleshy, stem fleshy in larger species, cartilaginous in small ones. Gills adnate, sinuate or decurrent, of varying colour in young, rust-brown when mature. Over 200 British species, all terrestrial.

107. *Rozites (= Pholiota) caperata (see* plate 47).

108. **Cortinarius elatior**   Cap leaden or yellow according to weather, more whitish along margin, convex, membraneous; stem lilac, downy; cortina slimy; gills ochre to lilac, then dark brown as spores mature, connected by veins. Woods. Common. Edible. Cap 4 in., stem $6\frac{1}{2}$ in.

109. **Cortinarius cinnamomeus** (from the colour)   Cap tawny to cinnamon, convex then more flattened with a central boss, silky and covered with yellowish fibrils; stem yellowish, fibrillose; cortina similar; gills yellowish then cinnamon, shiny, adnate, broad and crowded; flesh yellowish. Woods. Aug.–Feb. Common. Cap 3 in., stem $3\frac{1}{2}$ in.

110. **Cortinarius glaucopus** (Gk. *glaukos*, pale blue; *pous*, foot) (*see* plate 43).

111. **Cortinarius purpurascens** (L. *purpurascens*, turning purple— *i.e.*, the gills when bruised)   Cap brown to olive-brown, convex and depressed around margin which finally bends back, sticky; stem pale blue, darkens when touched, base bulbous; gills azure-blue then cinnamon, purple when bruised, rounded and crowded, curved towards stem; flesh blue. Woods. Sept.–Nov. Common. Edible. Cap 4 in., stem 3 in.

112. **Cortinarius violaceus** (from the colour)   Cap deep violet, even purple, fleshy, convex then flattened with a central boss, covered in fine scales, shiny; stem dark violet, stout and bulbous; cortina woolly, shining, turning rust from spore-dust; gills vivid violet, then rust from maturing spores, adnate, broad, firm and distant, joined by veins; flesh blue then white. Woods, under birch and beech. Aug.–Nov. Frequent. Edible. Cap and stem 4 in.

113. **Cortinarius collinitus** (L. *collinitus*, smeared) Cap orange-tawny, fleshy, convex then expanded with central boss, covered with slime, shiny when dry; stem violet or yellowish, firm, covered with a downy veil broken into scales; a ring may be present near apex; gills white or grey, then cinnamon, adnate, crowded; flesh white, more brownish in cap. Woods. July–Nov. Common. Edible. Cap 4¾ in., stem 5½ in.

114. **Cortinarius mucifluus** (L. *mucus*, slime; *fluere*, to flow) (*see* plate 44).

Genus **Inocybe** (Gk. *is*, *inos*, a fibre; *kube*, head—from the fibrous cap) Cap silky or fibrillose, the cuticle continuing across to the stem as a fibrillose cortina covering the gills in the young stage; stem mainly scaly, in some species smooth; gills sinuate, occasionally adnate or decurrent. Small in size. Terrestrial. Some poisonous. Over 50 British species.

115. **Inocybe scabra** (L. *scaber*, rough) Cap pale tan mixed with dark, flattened, spot-like scales, fleshy, conical; stem whitish, firm, covered in fibrils; gills whitish, then darken, adnexed, thin and crowded; flesh white. Conifer woods. June–July. Frequent. Cap and stem 1¾ in.

116. **Inocybe rimosa** (L. *rimosus*, cracked) Cap yellowish to date-brown, bell-shaped then flattened and turned back, surface cracked; stem whitish becoming brown-tinged, with fibrils at base; gills whitish then rust-coloured, free, with toothed edges. Smell earthy. Woods and open ground. June–Nov. Common. ☣Suspect. Cap 3 in., stem 3¼ in.

117. **Inocybe fastigiata** (L. *fastigare*, to slope) (*see* plate 45). ☠

118. **Inocybe geophylla** (Gk. *ge*, earth; *phullon*, leaf) (*see* plate 46). ☠

119. **Inocybe patouillardii** Cap at first entirely white, turning brick-red when bruised, dry and silky, convex with a central boss, margin becoming lobed and split, turning yellow to vermilion, finally brown; stem long, firm and solid, sometimes twisted; gills white, then grey to olive, crowded, adnexed; flesh firm and white, reddens when cut, spores brown. In grass along rides and woodland borders. Occasional. ☠Poisonous. Cap 3 in., stem 3–6 in.

Genus **Hebeloma** (Gk. *hebe*, youth; *loma*, a fringe—from the fringe-like veil) Similar to **Tricholoma** but with clay-coloured spores. Cap smooth or viscid, margin at first incurved; stem fleshy showing traces of veil as a ring; gills adnate or sinuate. Terrestrial. About 30 British species.

120. **Hebeloma fastibile** (L. *fastibilis*, unpleasant)   Cap pale tan, compact fleshy, more or less flattened with curved-under and wavy margin, smooth; stem solid, silky, white, covered in fibrils, may show traces of a ring; gills white then clay-coloured, broad, exuding drops; flesh white, tastes of radish, smells foetid. Usually in clusters. Woods. July–Nov. Common. 🟡Suspect. Cap 4 in., stem 2½ in.

Genus **Naucoria** (L. *naucum*, a small portion—from the almost absent veil)   Similar to **Collybia** but small and usually brownish. Cap fleshy, conical at first with an incurved margin, becoming flat; stem cartilaginous and hollow; gills free or adnate. About 50 British species.

121. **Naucoria horizontalis** (L. *horizontalis*, horizontal)   Pale cinnamon all over. Cap fleshy, convex and strongly curved; stem solid, short and smooth; gills rounded, free and broad. On branches, logs of elm. Late in year. Somewhat rare. Cap ¾ in., stem ½ in.

122. **Naucoria escharoides**   Cap tan then paler, convex then flat with curved-over margin; stem brownish, smooth but covered with fibrils; gills cinnamon, adnate, with a decurrent tooth. Under alder. Common. Cap 1 in., stem 1½ in.

15. Family Russulaceae

Genus **Russula** (L. *russulus*, reddish—from the colour of the cap in many species)   Does not exude milk. Cap fleshy, regular, becoming depressed; stem fleshy; gills adnate, rigid and brittle, make a rustling sound when stroked. Mainly terrestrial, in woods; many stand out in bright colours. About 60 British species, of which those with a mild taste are non-poisonous, others suspect.

123. **Russula nigricans** (L. *niger*, black)   Cap olive, then black, convex, then slightly depressed in centre; stem pale, then black; gills ochre, thick and distant, turning reddish if bruised; flesh white, reddens when broken, then turns black. Taste mild, then bitter. Woods. Common. Aug.–Nov. Edible. Cap 5½ in., stem 2½ in.
   The parasitic **Nyctalis asterophora** occurs on this species.

124. **Russula adusta** (L. *adustus*, scorched)   Cap whitish at first, then grey-brown, convex then depressed and slightly funnel-shaped; stem similar colour, swollen; gills white, adnate to decurrent, thin

and crowded; flesh white, then brown turning black, taste mild. Woods. Aug.–Nov. Edible. Cap 3½ in., stem 2¼ in.

The parasitic *Nyctalis parasitica* occurs on this species.

125. **Russula foetans** (L. *foetens*, stinking)   Cap dull yellow, fleshy, thin and rigid, expanded then depressed, sticky in wet weather, margin striated; stem whitish; gills whitish or straw-coloured, exuding drops when young, adnexed, crowded and joined by veins; flesh white then yellowish, taste bitter, smell strong. Woods. July–Nov. Common. ⊛Suspect. Cap 4 in., stem 3 in.

The parasitic *Nyctalis parasitica* occurs on this species.

126. **Russula cyanoxantha** (Gk. *kyanos*, blue; *xanthos*, yellow) (*see* plate 24).

127. **Russula integra** (L. *integer*, entire) (*see* plate 25).

128. **Russula drimeia**   Cap bright purple to dark red, fades with age, dull when dry, sticky when wet, margin incurved; stem has purple tinge; gills deep yellow or pale sulphur, adnexed, forked at base; flesh reddish, taste very bitter. Conifer woods. Common. ⊛Suspect. Cap 4 in., stem 3 in.

129. **Russula queletti** (*see* plate 26).

130. **Russula rubra** (= **atropurpurea**) (L. *rubra*, red)   Cap blood-red, almost black in centre, yellowish towards edge, convex then depressed or slightly flask-shaped, viscid and slightly wrinkled; stem white sometimes brownish or reddish; gills white to yellowish, free and rather crowded; flesh white may be stained brown, taste mild, a light and pleasant smell. Conifer woods. Aug.–Oct. Common. Cap 3½ in., stem 3 in.

131. **Russula emetica**   Sickener (Gk. *emetikos*, causing sickness) (*see* plate 27). ⊛

132. **Russula sanguinea** (L. *sanguineus*, bloody)   Cap blood-red, paling around the sharp-edged margin, fleshy and firm, convex then depressed becoming more funnel-shaped with an inner boss, smooth, moist after rain; stem reddish, firm and wrinkled, striated; gills white then cream, decurrent, crowded, narrow, joined by veins; flesh white and cheesy, reddish immediately under skin, taste bitter. Conifer woods. Aug.–Sept. Occasional. ⊛Suspect. Cap 4 in., stem 2¼ in.

133. **Russula ochroleuca** (Gk. *ochros*, pale yellow) Cap yellow turning pale, convex and smooth; stem ash-grey, slightly wrinkled; gills white, free, fragile; flesh yellowish under cap, taste acrid, smell pleasant. Woods. Common. Edible. Cap 3½ in., stem 2 in.

134. **Russula fragilis** Similar to **R. emetica** (plate 27) but smaller, paler in colour and with thinner cap; stem fragile and striate; gills thin and crowded; flesh is not red under cap. Woods. Common. Suspect. Cap 2¾ in., stem 2 in.

Genus **Lactarius** (L. *lac*, milk) Exudes a white milk when broken, the taste of which can be used as a means of identification in some. Cap fleshy; gills adnate or decurrent. Terrestrial, in woods. Mild tasting forms are edible. About 70 British species.

135. **Lactarius deliciosus** Saffron Milkcap (L. *deliciosus*, delicious) (*see* plate 18).

136. **Lactarius rufus** (L. *rufus*, red) Cap rufous, becoming depressed then funnel-shaped with a central boss, dry and smooth, margin curved-under when young; stem paler than cap, whitish, stuffed then hollow; gills pale ochre then rufous, decurrent, crowded, milk white; taste very bitter. Conifer woods. June–Dec. Common. Suspect. Cap 4 in., stem 2½ in.

137. **Lactarius quietus** Cap rich sienna, zoned, at first sticky then silky, opaque, margin curved-under at first; stem rufous with hairy base; gills white, then brick-coloured, slightly forked; milk white and sweet tasting, smell oily. Woods. Common. Edible. Cap 4 in., stem 3 in.

138. **Lactarius subdulcis** (L. *sub*, under; *dulcis*, sweet—referring to the delayed action of the taste) Cap rufous, remaining dark, at first convex then depressed, covered in small papillae; stem similar colour, covered in down; gills paler, crowded; flesh becoming reddish; milk white, sweet at first then bitter, smell pleasant. Woods. Common. Edible. Cap 3 in., stem 2 in.

139. **Lactarius piperatus** (L. *piperatus*, peppery) (*see* plate 22).

140. **Lactarius volemus** (named after *volema pira*, the Red Warden pear) (*see* plate 20).

141. **Lactarius mitissimus** (*see* plate 21).

142. **Lactarius torminosus** Woolly Milkcap (L. *torminosus*, griping) (*see* plate 19).

143. **Lactarius vellereus** (L. *vellus*, a fleece)   Entirely white. Cap downy, becoming stained yellow, at first convex then expanded and a little depressed, margin strongly incurved; stem yellowish, finely hairy; gills white then pale ochre, adnate, decurrent, finely haired, thick and branched; flesh white, yellow when broken; milk white, very acrid. Woods. Aug.–Dec. Common. Edible. Cap 8 in., stem 2 in.

144. **Lactarius fuliginosus** (L. *fuliginosus*, sooty)   Cap tan covered with a fine, coffee-coloured velvety down, later turning bright yellow with a central brown disc, at first convex then flattened; stem white then tan to brick-red, stuffed; gills white then yellow, decurrent, thin and branched, connected by veins; milk and flesh white then rose, finally saffron; taste mild then a little bitter. Woods and pastures. Aug.–Nov. Common. ☠Suspect. Cap 4½ in., stem 3 in.

16. Family Gomphidiaceae

Genus **Gomphidius** (Gk. *gomphos*, a large nail)   Similar in build to **Hygrophorus** but has black spores. Cap fleshy and viscid; stem somewhat scaly or cortinate; gills waxy, decurrent. Terrestrial, under conifers. Four British species.

145. **Gomphidius glutinosus** (L. *glutinosus*, sticky) (*see* plate 14).

17. Family Paxillaceae

Genus **Paxillus** (L. *paxillus*, a small peg—from the shape)   Cap fleshy with margin at first strongly incurved, stem central, lateral or absent. Gills decurrent, soft and easily separated from cap. Terrestrial. About 15 British species. This genus is sometimes divided into **Lepista** (white spores) and **Tapinia** (rust-coloured spores). The genus has affinities with **Boletus**.

146. **Paxillus involutus** (L. *involutus*, involuted) (*see* plate 12).

147. **Paxillus atrotomentosus** (L. *tomentum*, stuffing—*i.e.*, the downy covering) (*see* plate 13).

18. Family Boletaceae

Pored toadstools. The cap is fleshy, the stem central and often bulbous, occasionally ringed. Tubes are distinct from the flesh of the cap, staining blue in many species. In some species the pores are joined in compound openings and may also be toothed. The general build of the toadstool is robust. There are about 70 British species, all terrestrial, many edible. Many species form mycorrhizal links with tree-roots. This family is divided into a number of genera according to the length and shape of the tubes and the colour of the spores.

Genus *Boletus* (Gk. *bolos*, a clod—from the shape of the cap) Growth sturdy. Cap at first cushion-shaped and arched; stem stout and bulbous, finely netted above; spore-dust olive-brown.

148. *Boletus edulis* (L. *edulis*, eatable) Penny Bun (*see* plate 1).

149. *Boletus calopus* (*see* plate 3).

150. *Boletus satanas* Devil's Boletus (Gk. *Satanas*, the Devil— from its supposedly evil properties) Cap pale brown, turning white, rounded then convex, cushion-shaped, smooth and a little slimy; stem pink to dark blood-red or brownish, paling upwards to yellow, covered with a network of blood-red veins; tubes yellow and free; pores blood-red, then orange finally olive-brown, turning dark blue-green when touched, minute and round; flesh white then cream, turning blue if broken, reddish at base of stem. Woods on chalk. July–Oct. Uncommon. ⊛Suspect. Cap 6 in., stem 3 in.

151. *Boletus luridus* (L. *luridus*, sallow or dull) (*see* plate 4).

152. *Boletus versipellis* (L. *versipellis*, changeable) Cap rufous, rounded and cushion-shaped, dry, at first velvety then scaly, margin carries remnants of veil; stem long, whitish, rough and covered with small reddish scales when young, turning grey then black with age; tubes dull white, long; pores white then grey, minute and rounded; flesh white, greenish in stem. Woods, heaths and pastures. July– Nov. Common. Edible. Cap 4½ in., stem 6 in.

Genus *Tylopilus* Spore-dust and tubes a rosy red.

153. *Tylopilus (= Boletus) felleus* (L. *fel*, a gall—from the sharp taste) (*see* plate 2). ⊛

Genus *Leccinium* Flesh fibrous, stem slender and elongated, tubes long.

154. *Leccinium (= Boletus) scabrum* (L. *scaber*, rough) Cap light brown to dark brown or grey, occasionally orange, rounded and cushion-shaped, smooth, sticky when wet, finally rough, margin at first carries parts of veil; stem long, whitish or grey, striped and covered with minute grey scales which darken with age; tubes dull

white, free, long; pores minute and round; flesh white. Woods. May–Dec. Common. Edible. Cap $4\frac{1}{2}$ in., stem $7\frac{1}{2}$ in.

Genus *Xerocomus* Cap dry, velvety in young, slender stem without network markings. Spore dust yellow-brown.

155. *Xerocomus (= Boletus) badius* (*see* plate 6).

156. *Xerocomus (= Boletus) subtomentosus* (L. *sub*, slight; *tomentosus*, downy) Cap gold-brown to olive-brown, convex then cushion-shaped, soft, dry and downy, often cracked with yellow flesh showing through; stem bright yellow, streaked with crimson, firm, slender and ribbed; tubes golden yellow, adnate; pores large and angular; flesh yellowish, brown under stem turning a little blue above tubes, taste mild. Woods, heaths and pastures. July–Dec. Common. Edible. Cap $3\frac{1}{2}$ in., stem $3\frac{1}{2}$ in.

157. *Xerocomus (= Boletus) chrysenteron* (*see* plate 7).

158. *Xerocomus (= Boletus) parasiticus* Grows on the two earth-balls *Scleroderma aurantium* and *S. verrucosum* (*see* nos. 165 and 166). Cap yellow-brown, convex then flat, downy, often cracked; stem yellow and curved; tubes yellow then reddish, decurrent, short; pores compound. Frequent. Cap 2 in., stem 2 in.

Genus *Suillus* Cap slimy, tubes not decurrent, stem with or without a ring.

159. *Suillus (= Boletus) luteus* (L. *luteus*, yellow) (*see* plate 8).

160. *Suillus (= Boletus) variegatus* (L. *variegatus*, variable) (*see* plate 11).

161. *Suillus (= Boletus) granulatus* (L. *granulatus*, granulated) (*see* plate 10).

162. *Suillus (= Boletus) bovinus* (L. *bovinus*, ox-like—from the colour) Cap reddish yellow to rust-red, slimy, cushion-shaped then flattens, margin white, curled over beyond tubes; stem similar colour, a little swollen at the whitish base; tubes yellowish then green to dark olive-brown, shallow at first; pores toothed with compound opening; flesh whitish, reddish under skin of cap, blue over tubes. Common. Aug.–Nov. Edible. Cap $4\frac{1}{2}$ in., stem $2\frac{1}{2}$ in.

163. *Suillus (= Boletus) piperatus* (L. *piperatus*, peppery) Cap dull ochre-orange, convex and smooth, slightly sticky; stem similar colour, slender and fragile, arising from a yellow mycelium; base

contains yellow milk; tubes bright rust-red; pores large, angular, often toothed; flesh deep sulphur-orange, taste peppery. Woods and heaths. Aug.–Nov. Common. Edible. Cap 5½ in., stem 3 in.

Genus *Strobilomyces* (Gk. *strobilos*, a pine-cone; *mukes*, fungus) Differs from *Boletus* in the cap which is covered in overlapping scales, and having a ring and dark spores.

164. *Strobilomyces strobilaceus*   Cap white, becoming brownish or black, cushion-shaped then convex, skin broken into large, thick and woolly scales; stem similar colour, apex white, scaly below the thick, white, and woolly ring; tubes white, becoming brownish, long; pores white, becoming reddish when bruised; flesh white, reddish, then blackish. Woods. Occasional. Cap 6 in., stem 7 in. (Some authorities include this genus in a separate family, the Strobilo-mycetaceae.)

ORDER GASTERALES

19. Family Sclerodermataceae

Earth-balls. Spherical shaped fungi with a one-layered, thick peridium containing the gleba, firm at first until the skin breaks to release dark coloured clouds of spores. There is no capillitium.

Genus *Scleroderma* (Gk. *skleros*, hard; *derma*, skin) Fruit-body sessile or on a short base, wall firm and leathery, or corky, warted or scaly, or even smooth, breaks into irregular pieces, revealing the dark gleba which is traversed by sterile veins, and eventually powders. Terrestrial. About 5 British species.

165. *Scleroderma aurantium (= vulgare)*   Fruit-body white to yellowish, globular, sessile, with a thick wall covered in warts or close-fitting scales, attached to a dense mass of white, cord-like mycelium; gleba greyish white turning to purple-black, marbled with white, sterile veins. Ruptures by irregular slits of the peridium. Smell earthy. Open spaces near trees, sometimes attached to old tree-roots. July–Jan. Common. 4 in. across.

166. *Scleroderma verrucosum* (L. *verrucosus*, warted)   Fruit-body ochre to dull brown, globular, a little flattened, running into an elongated, stem-like base, sometimes sessile, covered in minute, dark warts or flattened scales, occasionally smooth; peridium thin and fragile above, breaking into cracks; gleba deep umber to purple-

brown, marbled with white veins, becoming powdery. Sandy ground. Jul.–Nov. Common. 3 in. tall, 5 in. across.

Both nos. 165 and 166 are parasitized by *Xerocomus (= Boletus) parasiticus* (*see* no. 158).

They are sometimes used in place of the real truffle in sausage-meat, *pâté de foie gras* and other foods.

20. Family Lycoperdaceae

Puff-balls. Rounded fungi with a double-walled peridium which opens at the top or irregularly, containing a powdery gleba, white and fleshy at first, together with a structure of elastic hyphae, the capillitium, which enables the fungus to "puff". Originally all British species were contained in the genus, but this is now reserved only for those species in which the spores are freed through an apical pore. Those species in which the peridium simply tears open belong to the genus *Calvatia*.

Genus *Lycoperdon* (Gk. *lukos*, a wolf; *perdon*, dung)   Outer wall, composed of spine-like tufts or warts, gradually disintegrates; inner wall smooth, dehisces by an apical pore to release the spores by a "puffing" action. Base sterile and usually stem-like.

167. *Lycoperdon gemmatum (=perlatum)* (L. *gemma*, a bud) (*see* plate 72).

168. *Lycoperdon piriforme*   Fruit-body white or grey to brownish, thin, soft and pear-shaped, attached at base to white, cord-like mycelia; outer peridium covered in minute spines and granules; inner peridium smooth; gleba white then greenish-yellow, finally deep brown; sterile base white at first, then discolours. In clusters on tree stumps. Common. Edible when young. $2\frac{1}{4}$ in. tall, $1\frac{3}{4}$ in. wide.

169. *Calvatia (=Lycoperdon) caelatum* (L. *caelatum*, engraved) Fruit-body white turning ochre, then brownish, oval; running into a short, stout, stem-like base attached to a thick mycelium. Outer peridium woolly covered with large warts and cracking into a net-work of patches. Warts later vanish; inner peridium fragile, falling to pieces to leave a cup-like base; gleba white to yellow, then olive, compact, forms half contents of peridium, separated from the sterile base by a diaphragm; spores dark olive, capillitium yellow. Woods, heaths and pastures. May–Nov. Common. Edible when young. $7\frac{1}{2}$ in. tall, $6\frac{1}{2}$ in. wide.

170. **Calvatia (= Lycoperdon) giganteum**   Giant Puff-ball. Fruit-body white then yellowish to brown, finally olive, globular but slightly pear-shaped, sessile and wrinkled at base, attached to cord-like mycelia; outer peridium at first downy then smooth and shiny, fragile, easily falling off; inner peridium brittle and soon breaks up; gleba white, then yellow, finally olive; sterile base almost absent; spores a deep olive-brown. Pastures and orchards. May–Nov. Occasional. Edible when young. 12 in. across. This is one of the largest British species of fungus.

21. Family Geasteraceae

Earth-stars. Puff-balls whose outer peridium splits and peels back on itself to a star shaped base on which it stands. The inner peridium is thereby exposed as the puff-ball. About 12 British species, found mainly on sandy soil.

Genus **Geastrum** (Gk. *ge*, earth; *aster*, a star)   As described above.

171. **Geastrum fimbriatum** (L. *fimbria*, a fringe)   Outer peridium ochre with white cracks, splitting into five or more segments which recurve below to form the star-like support for the globular sessile inner peridium. Former has a membraneous outer layer and a deep ochre, fleshy inner layer which soon cracks and peels off; mouth indistinct; gleba brown. Woods. Common. 2½ in. diameter.

172. **Geastrum fornicatum** (L. *fornicatus*, vaulted)   Outer peridium splits into four or five segments, the outer layer cup-shaped, brown and rough, remaining below ground; the inner layer, arched and attached to outer by its tips, is hard, thick and leathery, cracking and peeling in places; inner peridium dark brown to rust, pear-shaped and downy, with a ring-like swelling above the join to the short, white stem; mouth tubular. Meadows. Occasional. 6 in. tall, 3 in. across.

22. Family Nidulariaceae

Bird's-nest fungi. The gleba or spore-mass is contained in small egg-like bodies, called *peridiola*, a number of which are contained within the cup-like peridium which forms the "nest". Drops of rain help to remove the "eggs" which rupture to liberate the spores. In some cases the peridiola are attached to the wall of the peridium by a stalk or *funiculus*.

Genus **Nidularia** (L. *nidulus*, a little nest)   Fruit-body globular and sessile, with a single wall, at first closed then opening in an irregular manner to uncover numerous small peridiola which remain free of attachment. On twigs, leaves or soil. 3 British species.

173. **Nidularia pisiformis** (L. *pisum*, pea; *formis*, shape) Peridium slightly downy, whitish or brown, fixed to dead branch on a broad base. Peridiole brown and shiny, numerous and free. Occasional. $\frac{1}{4}$ in.

Genus **Cyathus** (Gk. *knathos*, a wine-cup) Fruit-body cylindrical, then bell-shaped, with a three-layered wall, closed at first by a diaphragm which later ruptures; peridiola attached to inner wall by a cord (funiculus). 2 British species.

174. **Cyathus striatus** (L. *striatus*, grooved) Peridium reddish brown or rust, rough and hairy on the outside, bell-shaped, the top at first curved in and closed by a diaphragm, then open; inside lead-coloured and strongly ridged or fluted; peridiola whitish. In groups on stumps. Common. $\frac{3}{4}$ in.

175. **Cyathus verrucosus** Peridium grey to ochre, silky, bell-shaped with wide opening tapering to narrow base; inside brownish and smooth; peridiola grey to black, circular and shiny, attached by a white funiculus. On bare ground, rotten wood. Common. $\frac{1}{4}$ in.

Genus **Crucibulum** (L. *crucibulum*, a crucible) Fruit-body globe-then cup-shaped, with a two-layered wall, the outer thick, the inner thin and stretching over the top as a diaphragm, then rupturing; peridiola numerous and attached to wall by a long cord. 1 British species.

176. **Crucibulum vulgare** Peridium grey or dull cinnamon, bell-shaped with a broad opening, slightly downy at first, then smooth; inside white, smooth and shiny; peridiola pale and circular. On twigs, etc. Common. $\frac{1}{4}$ in.

23. Family Phallaceae

Stink-horns. The gleba containing spores is slimy at first, and carried up when mature on a tall stem which emerges from a cup-like volva half-buried in the soil. The unpleasant smell in some attracts insects which feed on the gleba and help in distributing spores.

Genus **Phallus** (Gk. *phallos*, phallus) Cap net-like and conical, covered in a slimy gleba, and attached to the top of a hollow, cylindrical, spiny stalk; stalk emerges from the "egg", which remains at the base as a cup-shaped volva. Terrestrial. 2 British species.

177. **Phallus impudicus** Common Stink-horn (L. *impudicus*, shameless) Cap white, cylindrical, at first covered in green slime which has a network pattern; stalk white, elongated, spongy and cellular, expands rapidly from "egg"; volva white to yellowish, cup-like, attached to white cord-like mycelium; smell very pungent and foetid. Woods, gardens, usually among rotting tree-roots. May–Nov. Common. Up to 10 in. tall.

178. **Phallus imperialis** Similar to **P. impudicus**, but has a pinkish volva, pink or pale blue mycelium, a broader attachment of the cap, and a smell of liquorice. Sand-dunes, mainly. Rather rare. Up to 10 in. tall.

Genus **Mutinus** Differs from **Phallus** in having a narrower cap which is wholly attached to the stalk, and an ovoid volva.

179. **Mutinus caninus** Dog Stink-horn (L. *canis*, a dog) Cap reddish, grading into stalk, finger-shaped, wrinkled, at first covered with a green slime; stalk white with a rose tint, slender, cellular; volva white to yellowish, attached to white cord-like mycelium; smell slight. Old stumps and among leaves, often in hollows where water collects. June–Dec. Common. Up to 5 in. tall.

ORDER TREMELLALES

24. Family Tremellaceae

Jelly-fungi with longitudinally septate basidia.

Genus **Tremella** (L. *tremere*, to tremble) Fruit-body gelatinous or waxy, lobed or brain-like, the hymenium spread over the whole surface. Usually on wood. About 15 British species.

180. **Tremella mesenterica** (Gk. *mesos*, middle; *enteron*, intestine—from its appearance) Fruit-body bright orange with brain-like surface, powdered with spore-dust; flesh similar colour, gelatinous, becoming firm and tough. Dead branches. Jan.–Dec. Common. 3 in. diameter.

Genus **Exidia** (Gk. *exidio*, to exude) Fruit-body gelatinous, oval with wavy surface, sessile or attached by short stem, cup-shaped when young, sterile on upper surface. Hymenium is inferior, smooth and veined. On wood. Two British species.

181. *Exidia recisa* (L. *recisus*, cut off)   Fruit-body amber-brown, surface wavy, rough and dotted; on a small stem; hymenium more brightly coloured, plane but veined; flesh similar colour, gelatinous, soft. On branches, especially willow. Sept.–Dec. Common 1¼ in. diameter.

25. Family Dacrymycetaceae

Jelly-fungi with divided basidia.

Genus **Calocera** (Gk. *kalos*, beautiful; *keras*, horn)   Formerly placed in family Clavariaceae. Differs in being gelatinous, hardening when dry, and having spores which become septate on germination. Fruit-body erect, simple or branched; hymenium smooth. About 7 British species, all on wood.

182. *Calocera viscosa* (L. *viscosus*, sticky)   Fruit-body a golden egg-yolk yellow, turning orange when dry, branched, viscid, linear and long-rooted; branches a similar colour, straight and generally forked. On conifer stumps. Common. 2½ in. tall.

26. Family Auriculariaceae

Genus *Auricularia* (L. *auricularia*, an ear)   Fruit-body gelatinous when moist, cartilaginous when dry, rounded and more or less cup-shaped; broadly attached; hymenium smooth or ribbed.

183. *Auricularia mesenterica* (Gk. *mesos*, middle; *enteron*, intestine —from its appearance)   Fruit-body reddish, tawny or grey, circular and cup-shaped, surface lobed and zoned; hymenium above, grey to brownish olive, dusted with spores; flesh reddish, gelatinous, then tough. Stumps and felled trees, especially elm, often in tiers. Jan.–Dec. Common. 8 in. diameter.

184. *Auricularia auriculajudae* Jew's Ear or Judas Ear (L. *auricula*, ear; *Juda*, a Jew)   Fruit-body grey, then reddish brown or olive, finally black, cup- or ear-shaped, downy and transparent, surface in folds; hymenium pale to greyish, then reddish brown, veined; flesh whitish, gelatinous, then tough. On trunks, especially elder. The elder is sometimes quoted as the tree in which Judas hanged himself. Jan.–Dec. Common. Edible. 3 in. diameter.
   Formerly used in medicine for treating dropsy and sore throats.

CLASS ASCOMYCETES

ORDER PEZIZALES (= Discales) (formerly Discomycetes)

27. Family Pezizaceae

Pixie-cups, Elf-cups. Fruit-body cup-shaped, spoon-shaped, sessile or borne on a stalk, at first closed then opening to reveal the hymenium on the upper surface.

Genus *Peziza* Fruit-body at first closed then expands into a cup-shape or disc, usually sessile, fleshy; hymenium superior, lining upper surface, smooth, granulated or veined.

185. ***Peziza vesiculosa*** (L. *vesiculosus*, bladder-like) Cup pale, yellowish with a reddish tinge, globular at first with small opening, spreading out into more irregular bell-shape with notched rim, outside whitish, granular, sessile, fleshy and transparent. In clusters in gardens, manure-beds, on rotten leaves, etc. Spring–autumn. 2 in. diameter.

186. ***Peziza cochleata*** (L. *cochleatus*, spiral-shape) Cup a deep brown, externally white to dull ochre, globular then expanded with turned-in margin, sessile, sometimes split to base or contorted and folded. On ground, summer and autumn. Common. Edible. 3 in. diameter.

Genus *Aleuria* (Gk. *aleuron*, flour, from the texture of the surface)

187. ***Aleuria aurantia*** Orange-peel fungus (L. *aurantius*, orange-colour) Cup shiny, reddish orange, paler outside, white when dry, downy, cup-shaped, sessile; margin more or less bent inwards, thin and soft. In clusters on ground. Summer and autumn. Common. Edible. 4 in. diameter.

Genus *Otidea* (Gk. *otion*, a small ear) Fruit-body ear-shaped, short-stalked, obliquely contorted and incised down one side; outer surface downy.

188. ***Otidea leporina*** (L. *leporinus*, hare-like) Cup cinnamon to rust, paler on outside, generally smooth, cup- or ear-shaped with thin and wavy involute margin; base stem-like and short, fleshy and tough. On dead leaves. Autumn. 3 in. tall, 2 in. wide.

28. Family Helvellaceae

Morels. Fruit-body lobed, support on a distinct stalk.

Genus **Helvella**   Fruit-body a cap-like structure with thin, irregular and drooping lobes, saddle-shaped, supported at centre by an elongated, often ribbed stem; hymenium on upper surface.

189. **Helvella crispa** (L. *crispus*, wrinkled)   False Morel (*see* plate 77).

Genus **Gyromitra** (Gk. *gyros*, round; *mitra*, a turban)   Similar to *Morchella* (*see below*) but the hymenial layer has a brain-like surface of ribs and folds.

190. **Gyromitra esculenta** (L. *esculentus*, edible)   Cap tawny, turning to chestnut or coffee colour with a violet sheen, swollen with irregular folds, brain-like in appearance, hollow and thin, white and smooth underneath; stem whitish, tinged reddish or purple, grooved at base and apex, hollow above, brittle. Pine woods, on sand in spring. Uncommon.   ⊛Suspect, in spite of name. Cap 3 in., stem 3 in.

29. Family Morchellaceae

True Morels. Caps with a varying degree of roundness and a honeycombed surface; the hymenium within the hollows, the ribs sterile, on a ribbed stalk. Edible.

Genus **Morchella** (German *morchel*, a morel)   Fruit-body stalked, with a conical, club- or globe-shaped cap bearing a deeply folded surface, the hymenial layer, in the hollows.

191. **Morchella esculenta** (L. *esculentus*, edible)   Round Morel (*see* plate 78).

192. **Morchella conica** (L. *conicus*, conical) (*see* plate 79).

ORDER HELOTIALES

30. Family Geoglossaceae

Earth-tongues. Fruit-body club-, tongue- or spatula-shaped, containing the hymenium; stalked; terrestrial.

Genus **Geoglossum** (Gk. *ge*, earth; *glossa*, tongue)   Fruit-body black or deep green, fleshy, erect, spoon- or club-shaped, stalked.

193. **Geoglossum hirsutum** (L. *hirsutus*, hairy)   Fruit-body rounded or lance-shaped with black, hairy, hymenial layer, on a long stalk. In clusters, in damp meadows. Frequent. Cap $\frac{1}{2}$ in., stalk 3 in.

31. Family Helotiaceae

The fruit-body is shaped like a top or a bowl. It is gelatinous, and often very small. The spores are dark brown.

Genus **Bulgaria** (L. *bulga*, a leather bag)   Fruit-body gelatinous, sessile, smooth, usually cup-shaped; spores dark brown.

194. **Bulgaria inquinans** (L. *inquinans*, staining)   Fruit-body deep umber, outside wrinkled and rough, inside black and shiny, at first closed then opening into a cup shape on a short stem; flesh brown and rubbery. In clusters on old trunks, especially beech. Dark spores leave a sooty deposit on surrounding wood. Common. $1\frac{1}{2}$ in. diameter.

ORDER TUBERALES

32. Family Eutuberaceae

Truffles. Fruit-body roughly globe-shaped, flesh leathery and firm with a warted surface. Veins inside give a marbled appearance.

Genus **Tuber** (L. *tuber*, a swelling)   Characters as for family.

195. **Tuber aestivum** (L. *aestivus*, summer-time)   Fruit-body shiny blue-black, more brownish when dry, globe-shaped and a little depressed, tending to look four-sided, surface irregular covered with hard, lumpy warts which later split; inside white, then yellowish, brown when ripe, containing many branched veins; slight, pleasant smell. Below ground, especially in beech woods. Autumn and winter. 4 in. diameter.

This is the best species of truffle in Britain. The French species **T. melanospermum** is mainly used in *pâté de foie gras*.

33. Family Terfeziaceae

The fruit-body is dull yellow and smooth, rarely black or warted, pale to brown inside and veined. It is usually half-buried in soil.

Genus **Choiromyces** (Gk. *choiros*, pig; *mukes*, fungus)   Fruit-body globe-shaped with smooth, naked surface; flesh white and lined with numerous veins.

196. **Choiromyces meandriformis** (L. *meander*, winding; *forma*, shape) White Truffle (*see* plate 80).

ORDER PLECTOASCALES

34. Family Elaphomycetaceae

Fruit-body below ground, with a crusty skin, powdery inside.

Genus **Elaphomyces** (Gk. *elaphos*, stag; *mukes*, fungus) Fruit-body globe-shaped, a little flattened with irregular surface covered in mycelium. Wall thick in two layers, inside filled with a scattered, powdery mass of dark-coloured spores containing hyphae (capillitium). May form mycorrhiza with roots of pine

197. **Elaphomyces granulatus** (L. *granum*, grain) Fruit-body yellowish or reddish brown, covered with rounded warts, and with mycelium when young; spore-mass black when ripe lined with greyish veins; smells of garlic. Below ground in woods, especially beech. Not uncommon. 1½ in. diameter. Is sometimes parasitized by *Cordyceps capitata* (*see* no. 201).

ORDER SPHAERIALES

35. Family Sphaeriaceae

Fruit-body black, resembling charcoal in texture, club-shaped, hard, with asci borne inside perithecia.

Genus **Xylaria** (Gk. *xylon*, wood) Fruit-body black, erect, cylindrical, club-shaped or thread-like, simple or branched, corky or leathery. Perithecia in upper half with mouths projecting from surface giving it a rough appearance. On or near wood.

198. **Xylaria polymorpha** (Gk. *polymorphos*, many forms) Fruit-body first whitish, then grey, finally black, solitary or in bunches, club-shaped or cylindrical, varying much in size; head rough surfaced with projecting mouths of perithecia. Gregarious, on old stumps. Jan.–Dec. Common. 2 in. tall.

199. **Xylaria hypoxylon** Candle-snuff Fungus (Gk. *hypo*, below; *xylon*, wood) Fruit-body erect, simple or branched, lanceolate, head white at first, turning black, rough from the crowded perithecia; apex sterile; stem short and downy. On old stumps. Common. 2½ in. tall.

This fungus is sometimes luminous, depending on the condition of the wood.

ORDER CLAVICIPITALES (formerly Pyrenomycetes)

36. Family Hypocreaceae

Small, parasitic fungi with club-shaped fruit-body and elongated stem attached to insect larvae or underground fungi—*i.e.*, truffles.

Genus **Cordyceps** (Gk. *kordule*, a club) Stroma erect, fleshy and stalked with a fertile, club-shaped head in which are buried the perithecia. Grows mainly on dead insects. Once believed to be insects which changed themselves into plants, described as "vegetable caterpillars".

200. **Cordyceps militaris** (L. *militaris*, soldier-like) Stroma yellowish red or crimson, club-shaped, narrow at both ends, covered in tubercles due to protruding openings of the perithecia, single or in clusters. Grows on caterpillars or pupae in damp woods. Autumn. Common. 2 in. long.

201. **Cordyceps capitata** (L. *capitatis*, with a head) Stroma oval, yellow or rust-brown to black, rough from projecting mouths of the perithecia; stem yellow, smooth, then darkens. Parasitic on the truffle **Elaphomyces granulatus** (no. 198). Woods. Not uncommon. 4 in. long.

Index

ALPHABETICAL LIST OF SCIENTIFIC NAMES

Agaricaceae 210
Agaricus arvensis 22, 29, 108, 212
Agaricus bisporus 212
Agaricus bisporus (var. *albida*) 22, 212
Agaricus campestris 22, 29, 104, 106, 212
Agaricus haemorrhoidarius 104
Agaricus perrarus 102, 104
Agaricus silvaticus 104, 212
Agaricus silvicola 88, 90, 106, 108, 212
Agaricus xanthodermus 106, 108, 212
Aleuria aurantia 178, 180, 231
Amanita caesarea 94
Amanita citrina (=*mappa*) 29, 88, 92
Amanita gemmata 92
Amanita mappa (see *Amanita citrina*)
Amanita muscaria 17, 25, 92, 94, 98, 210
Amanita pantherina 94, 96, 98, 210
Amanita phalloides 24, 29, 88, 210
Amanita rubescens 29, 96, 98, 210
Amanita spissa (=*excelsa*) 96, 210
Amanita verna 24, 90, 210
Amanita virosa 24, 88, 90, 210
Amanitopsis fulva 29, 211
Amanitopsis vaginata 29, 211
Armillariella (=*Armillaria*) *mellea* 29, 138, 206
Auriculariaceae 230
Auricularia (=*Hirneolea*) *auricula-judae* 30, 230
Auricularia mesenterica 30, 230

Bolbitius fragilis 216
Bolbitius vitellinus 29
Boletaceae 222
Boletinus cavipes 48
Boletus aeruginascens (see *Suillus*)

Boletus badius (see *Xerocomus*)
Boletus bovinus (see *Suillus*)
Boletus calopus 36, 223
Boletus chrysenteron (see *Xerocomus*)
Boletus edulis 22, 29, 32, 94, 223
Boletus erythropus 38
Boletus felleus (see *Tylopilus*)
Boletus granulatus (see *Suillus*)
Boletus grevillei (see *Suillus*)
Boletus luridus 36, 38, 223
Boletus luteus (see *Suillus*)
Boletus parasiticus (see *Xerocomus*)
Boletus piperatus (see *Suillus*)
Boletus purpureus 38
Boletus rufescens (see *Leccinium*)
Boletus satanas 38, 223
Boletus scaber (see *Leccinium*)
Boletus subtomentosus (see *Xerocomus*)
Boletus variegatus (see *Suillus*)
Boletus versipellis 223
Bovista nigrescens 176
Bulgaria inquinans 30, 233

Calocera viscosa 164, 230
Calvatia (=*Lycoperdon*) *vaelata* 176, 226
Calvatia (=*Lycoperdon*) *gigantea* 22, 176, 226
Cantharellaceae 203
Cantharellus cibarius 22, 29, 94, 154, 203
Cantharellus cibarius (var. *pallidus*) 70, 156
Cantharellus cinereus 162
Cantharellus infundibuliformis 203
Cantharellus lutescens 158
Cantharellus tubaeformis 158, 203
Choiromyces meandriformis 190, 234
Clavariaceae 202

Clavaria aurea 164, 166
Clavaria cinerea 201, 202
Clavaria (= *Clavulina*) *cristata* 202
Clavaria flava 164, 166
Clavaria formosa 164, 166
Clavaria fusiformis 202
Clavaria pallida 164, 166
Clavaria pistillaris 74
Clavaria (= *Ramaria*) *stricta* 202
Clavaria truncata 160
Clavaria vermicularis 202
Clavulina cristata (see *Clavaria*)
Clitocybe alexandri 128, 130
Clitocybe aurantiaca (see *Hygrophoropsis*)
Clitocybe candida 130
Clitocybe clavipes 205
Clitocybe dealbata 60, 132, 134, 206
Clitocybe flaccida 206
Clitocybe fragrans 206
Clitocybe geotropa 130, 206
Clitocybe infundibuliformis 130, 205
Clitocybe nebularis 126, 128, 150, 205
Clitocybe odora 206
Clitocybe rivulosa 132
Clitopilus prunulus 132
Collybia fusipes 207
Collybia maculata 29, 207
Collybia radicata 205
Collybia velutipes 205
Coprinaceae 213
Coprinus atramentarius 86, 213
Coprinus comatus 22, 86, 213
Coprinus micaceus 213
Coprinus plicatilis 213
Cordyceps capitata 234, 235
Cordyceps militaris 235
Cortinarius cinnamomeus 217
Cortinarius collinitus 118, 218
Cortinarius elatior 118, 217
Cortinarius glaucopus 104, 116, 217
Cortinarius mucifluus 118, 218
Cortinarius mucosus 118
Cortinarius purpurascens 217
Cortinarius trivialis 118
Cortinarius violaceus 217
Cortinariaceae 217
Cratarellus cornucopioides 29, 70, 74, 162, 203
Crucibulum vulgare 228

Cyathus striatus 228
Cyathus verrucosus 228

Dacrymycetaceae 230
Daedalea quercina 30, 200

Elaphomyces granulatus 234, 235
Elaphomycetaceae 234
Entoloma lividum (see *Rhodophyllus sinuatus*)
Eutuberaceae 233
Exidia recisa 230

Fistulinaceae 201
Fistulina hepatica 29, 201
Fomes applanatum (see *Ganoderma*)

Ganoderma (= *Polyporus*) *lucidum* 200
Ganoderma (= *Fomes*) *applanatum* 29, 200
Geastraceae 227
Geastrum fimbriatum 227
Geastrum fornicatum 227
Geoglossaceae 232
Geoglossum hirsutum 233
Gomphidiaceae 222
Gomphidius glutinosus 46, 52, 222
Gomphidius roseus 52
Gomphidius rutilus 58
Gomphus clavatus 160
Guepinia helvelloides 160
Gyromitra esculenta 182, 184, 232
Gyromitra gigas 182
Gyromitra infula 182

Hebeloma fastibile 219
Helotiaceae 232
Helvellaceae 232
Helvella crispa 184, 232
Helvella elastica 182
Helvella lacunosa 184
Hydnaceae 198
Hydnum erinaceus 198
Hydnum repandum 29, 74, 168, 198
Hygrophoraceae 204
Hygrophoropsis (= *Clitocybe*) *aurantiaca* 154, 204
Hygrophorus agathosmus 62
Hygrophorus chrysaspis 60
Hygrophorus chrysodon 60

Hygrophorus conicus 64, 204
Hygrophorus cossus 60
Hygrophorus hypothejus 62, 204
Hygrophorus lucorum 62
Hygrophorus olivaceoalbus 62
Hygrophorus penarius 60
Hygrophorus piceae 60
Hygrophorus pratensis 204
Hygrophorus psittacinus 64
Hygrophorus puniceus 64
Hypholoma capnoides (see Nema-
    toloma)
Hypholoma fasciculare (see Nema-
    toloma)
Hypocreaceae 235

Inocybe fastigiata 120, 146, 218
Inocybe geophylla 122, 218
Inocybe patouillardii 116, 122, 218
Inocybe praetervisa 120
Inocybe rimosa 218
Inocybe scabra 218

Laccaria amethystina 29, 136, 207
Laccaria laccata 136, 206
Lactarius deliciosus var. semi-
    sanguifluus 22, 29, 66, 221
Lactarius fuliginosus 221
Lactarius mitissimus 70, 72, 221
Lactarius piperatus 70, 74, 213, 221
Lactarius porninsis 72
Lactarius quietus 92, 221
Lactarius rufus 70, 221
Lactarius sanguifluus 66
Lactarius subdulcis 221
Lactarius torminosus 68, 221
Lactarius vellereus 74, 209, 221
Lactarius volemus 70, 72, 221
Lactarius zonarius 68
Leccinium carpini 40
Leccinium (= Boletus) rufescens 40
Leccinium (= Boletus) scabrum 29,
    40, 223
Lentinaceae 209
Lepiota brunneo-incarnata 102
Lepiota granulosa 211
Lepiota procera 22, 29, 100, 102,
    211
Lepiota rhacodes 100, 102, 211
Leptoporus albidus 172
Lycoperdaceae 226

Lycoperdon caelatum (see Calvatia)
Lycoperdon giganteum (see Calvatia)
Lycoperdon hiemale 176
Lycoperdon perlatum (= gemmatum)
    29, 174, 226
Lycoperdon pyriforme 174, 226

Marasmius oreades 20, 22, 29, 134,
    208
Marasmius perforans 134
Marasmius peronatus 208
Marasmius scorodonius 134, 136
Merulius lacrymans 30, 198
Morchellaceae 232
Morchella conica 186, 188, 232
Morchella deliciosa 108
Morchella elata 108
Morchella esculenta 106, 188, 232
Mutinus caninus 229
Mycena polygramma 208
Mycena pura 136, 208

Naucoria escharoides 219
Naucoria horizontalis 219
Nematoloma (= Hypholoma) cap-
    noides 110, 112, 114, 216
Nematoloma (= Hypholoma) fas-
    ciculare 30, 110, 112, 216
Nematoloma sublateritium 112
Nidulariaceae 227
Nidularia pisiformis 228
Nyctalis asterophora 209
Nyctalis parasitica 209

Omphalia pyxidata 208
Omphalia umbellifera 209
Otidea leporina 180, 231
Otidea onotica 180
Oudemansiella (= Armillaria)
    mucida 206

Panaeolus campanulatus 215
Panus conchatus 160
Paxillaceae 222
Paxillus atrotomentosus 54, 222
Paxillus involutus 29, 54, 56, 124,
    222
Penicillium notatum 21
Pezizaceae 231
Peziza cochleata 231
Peziza vesiculosa 231

Phaeolus (= *Polyporus*) *schweinitzii*
  200
Phallaceae 228
Phallus imperialis 229
Phallus impudicus 229
Pholiota caperata (see *Rozites*)
Pholiota carbonaria 29
Pholiota mutabilis 112, 114, 138
Pholiota spectabilis 216
Pholiota squarrosa 30, 114, 216
Piptoporus (= *Polyporus*) *betulinus*
  29, 199
Pleurotus ostreatus 22, 209
Pleurotus ulmarius 209
Plicariella fulgens 178, 180
Pluteus cervinus 211
Polyporaceae 198
Polyporus betulinus (see *Piptoporus*)
Polyporus caecius 198
Polyporus lucidus (see *Ganoderma*)
Polyporus squamosus 29, 199
Polyporus sulphureus 29, 199
Polystictus radiatus 201
Polystictus versicolor 29, 201
Protohydnaceae 195
Psathyra corrugia 214
Psathyra fatua 214
Psathyrella disseminata 214
Psathyrella gracilis 214
Psilocybe semilanceata 216

Ramaria stricta (see *Clavaria*)
Rhodopaxillus (= *Tricholoma*)
  nudus 22, 29, 46, 150, 205
Rhodopaxillus panaeolus 152
Rhodopaxillus (= *Tricholoma*)
  personatus 22, 128, 150, 152, 205
Rhodophyllaceae 213
Rhodophyllus nidorosus 126
Rhodophyllus rhodopolius 126
Rhodophyllus sinuatus (= *Entoloma*
  lividum) 126 128, 130, 213
Rozites (= *Pholiota*) *caperata*
  76, 92, 118, 124, 217
Russulaceae 219
Russula adusta 76, 209, 219
Russula alutacea 80
Russula badia 80
Russula cyanoxantha 70, 78, 220
Russula delica 74

Russula emetica 29, 74, 84, 220
Russula foetans 220
Russula firmula 80
Russula fragilis 84, 220, 221
Russula integra 80, 220
Russula leuteotacta 84
Russula nigricans 209, 219
Russula ochroleuca 29, 221
Russula olivacea 80
Russula queletti 82, 220
Russula rosea 74, 84
Russula rubra (= *atropurpurea*) 220
Russula sanguinea 82, 220
Russula sardonia 82
Russula vesca 78
Russula xerampelina 82

Sarcodon amarescens 170
Sarcodon (= *Hydnum*) *imbricatus*
  170, 198
Sarcoscypha coccinea 178
Sclerodermataceae 225
Scleroderma aurantium (= *vulgare*)
  29, 190, 224, 225
Scleroderma verrucosum 224, 225
Sparassis crispa 203
Sphaeriaceae 234
Stereum hirsutum 30, 197
Stereum purpureum 29, 197
Strobilomyces strobilaceus 225
Strophariaceae 215
Stropharia aeruginosa 215
Suillus (= *Boletus*) *aeruginascens*
  48, 66
Suillus (= *Boletus*) *bovinus* 52
Suillus (= *Boletus*) *granulatus* 46,
  50, 52, 224
Suillus (= *Boletus*) *grevillei* 46, 48
Suillus (= *Boletus*) *luteus* 46, 50, 52,
  58, 180, 224
Suillus (= *Boletus*) *piperatus* 94. 224
Suillus (= *Boletus*) *tridentinus* 48
Suillus (= *Boletus*) *variegatus* 52,
  224

Terfeziaceae 233
Thelepora terrestris (= *laciniata*)
  203
Trametes gibbosa 30, 172, 201

Trametes hirsuta 172
Trametes rubescens 201
Tremellaceae 229
Tremella mesenterica 229
Tricholomataceae 204
Tricholoma cingulatum 144
Tricholoma flavovirens 88, 146, 148
Tricholoma gambosum 22, 29, 122, 126, 204
Tricholoma inamoenum 148
Tricholoma nudum (see *Rhodopaxillus*)
Tricholoma orirubens 142, 144
Tricholoma pardinum 140, 144
Tricholoma personatum (see *Rhodopaxillus*)
Tricholoma rutilans 205
Tricholoma scalpturatum 140, 144
Tricholoma sejunctum 146
Tricholoma sulphureum 146, 148, 205
Tricholoma terreum 140, 142, 144, 204

Tuber aestivum 23, 233
Tylopilus (= *Boletus*) *felleus* 32, 34, 223

Ungulina (= *Fomes*) *ulmaria* 201

Volvariella (= *Volvaria*) *volvacea* 213

Xerocomus (= *Boletus*) *badius* 29, 34, 42, 54, 92, 124, 224
Xerocomus (= *Boletus*) *bovinus* 29
Xerocomus (= *Boletus*) *chrysenteron* 36, 44, 224
Xerocomus (= *Boletus*) *parasiticus* 224
Xerocomus (= *Boletus*) *subtomentosus* 34, 44, 224
Xylaria hypoxylon 234
Xylaria polymorpha 234